W0230221

Biotreatment of Industrial Effluents

Biotreatment of Industrial Effluents

Editor

Sudhir Deshpande

scitus

Biotreatment of Industrial Effluents

Edited by **Sudhir Deshpande**

Printed in 2017

ISBN: 978-1-68117-337-5

Library of Congress Control Number: 2015939249

© 2016 by
SCITUS Academics LLC,
616, Corporate Way, Suite 2, 4766,
Valley Cottage, NY 10989

www.scitusacademics.com

This book contains information obtained from highly regarded resources. Copyright for individual articles remains with the authors as indicated. All chapters are distributed under the terms of the Creative Commons Attribution License, which permits unrestricted use, distribution, and reproduction in any medium, provided the original author and source are credited.

Notice

Reasonable efforts have been made to publish reliable data and views articulated in the chapters are those of the individual contributors, and not necessarily those of the editors or publishers. Editors or publishers are not responsible for the accuracy of the information in the published chapters or consequences of their use. The publisher believes no responsibility for any damage or grievance to the persons or property arising out of the use of any materials, instructions, methods or thoughts in the book. The editors and the publisher have attempted to trace the copyright holders of all material reproduced in this publication and apologize to copyright holders if permission has not been obtained. If any copyright holder has not been acknowledged, please write to us so we may rectify.

Contents

vi

Preface

Biotreatment of industrial effluents describes difficulties encountered in biological treatment of wastewater with highly variable influent characteristics. Typical design aspects of biological processes are presented and discussed with respect to their success in treating highly fluctuating wastewaters. In general, biomass retention is a key factor for dealing with highly fluctuating and/or inhibitory wastewater, but the how it operates also affects the stability of performance, as it was shown that dynamic operation instead of operation at a constant flow enhances biodegradation onset and more evenly distributed activity. Biotreatment of industrial wastewater is often challenged by operation under transient states with respect to organic loads, pollutants, and physical characteristics. Furthermore, the potential presence of inhibitory compounds requires careful monitoring and adequate process design.

Editor

Recent Applications of Electrocoagulation in Treatment of Water and Wastewater—A Review

Ville Kuokkanen[1], Toivo Kuokkanen[1], Jaakko Rämö[2], and Ulla Lassi[1,3]

[1]Department of Chemistry, University of Oulu, Oulu, Finland
[2]Thule Institute, University of Oulu, Oulu, Finland
[3]Kokkola University Consortium Chydenius, Kokkola, Finland

ABSTRACT

During the last two decades, and particularly during the last few years, the environmental sector has shown a largely growing interest in the treatment of different types of water and wastewater by electrocoagulation (EC). The aim of this work was to review studies, conducted mainly during 2008-2011, on the wide and versatile range of feasible EC applications employed in the purification of different types of water and wastewater. The EC applications discussed here were divided into 7 following categories: tannery, textile and colored wastewater; pulp and paper industry wastewater; oily wastewater;

food industry wastewater; other types of industrial wastewater; surface water as well as model water and wastewater containing heavy metals, nutrients, cyanide and other elements and ions. In addition, this paper presents an overview of the optimum process conditions (treatment times, current densities, and initial pH) and removal efficiencies (mostly high) achieved for the EC applications discussed. In the vast majority of the studies discussed in this review, the aforementioned values were found to be in the range of 5 - 60 min (typically less than 30 min), 10 - 150 A/m^2 and near neutral pH, respectively. Both operating costs and electrical energy consumption values were found to vary greatly depending on the type of solution being treated, being between 0.0047 - 6.74 EUR/m^3 and 0.002 - 58.0 kWh/m^3, but in general they were rather low (typically around 0.1 - 1.0 EUR/m^3 and 0.4 - 4.0 kWh/m^3).

INTRODUCTION

Electrocoagulation (EC) is an emerging technology that combines the functions and advantages of conventional coagulation, flotation, and electrochemistry in water and wastewater treatment. Each of these fundamental technologies has been widely studied separately. However, a quantitative appreciation of the mechanism of interaction between these technologies employed in an electrocoagulation system is absent [1].

EC has been known for over a century. Aluminium/ iron-based electrocoagulation was patented in the US already in 1909. EC was studied extensively in the latter half of the 20th century in both the US and the Soviet Union (former USSR), but at that time it was not found to be widely feasible for water treatment. This was mainly due to the then high electricity and investment costs [2].

Meanwhile, the demand for quality drinking water quality is increasing globally and environmental regulations regarding wastewater discharge are becoming increasingly stringent. Therefore, it has become necessary to develop more effective treatment methods for water purification and/or enhance the operation of current methods. This and eco-friendliness have led to increasing global interest in electrocoagulation as a research subject. Over the course of the last few decades, literature in the environmental sector has indeed shown a growing interest towards the treatment of different types of wastewater by EC. Particularly during the last few years, the amount of published literature on EC applications seems to have increased substantially.

Practical review papers on EC applications have been largely absent so far. To the best of our knowledge, only a few authors, e.g. [3,4] have addressed the subject recently, in addition to older reviews, (namely [1,2,5]), even though a significant number of studies on EC have been reported in the literature since then. Therefore, there is a need for an update on recent applications of EC. The aim of this work was to accomplish this, and based on the literature, to present an overview of practical optimum treatment times, current densities, electricity consumption, and operating costs in a wide and versatile range of feasible applications of EC in water and wastewater treatment, studied mainly during the years 2008- 2011.

BACKGROUND

Principles of Electrocoagulation

Electrolysis is a process in which oxidation and reduction reactions take place when electric current is applied to an electrolytic solution. Electrocoagulation is based on dissolution of the electrode material used as an anode. This so-called "sacrificial anode" produces metal ions which act as coagulant agents in the aqueous solution in situ [1]. At its simplest, an electrocoagulation system consists of an anode and a cathode made of metal plates, both submerged in the aqueous solution being treated [3]. The electrodes are usually made of aluminum, iron, or stainless steel (SS), because these metals are cheap, readily available, proven effective, and non-toxic. Thus they have been adopted as the main electrode materials used in EC systems [6,7]. The configurations of EC systems vary. An EC system may contain either one or multiple anode-cathode pairs and may be connected in either a monopolar or a bipolar mode [3]. During EC, the following main reactions take place at the electrodes. Anodic reactions [2]:

$$Al(s) \rightarrow Al^{3+} + 3e^- \qquad E^0 = +1.66 \text{ V}$$

$$(1)$$

$$Fe(s) \rightarrow Fe^{2+} + 2e^- \quad E^0 = +0.44 \text{ V} \tag{2}$$

$$2H_2O(l) \rightarrow O_2(g) + 4H^+ + 4e^- \quad E^0 = -1.23V \tag{3}$$

Ferrous iron may be oxidized to Fe^{3+} by atmospheric oxygen or anode oxidation, and may be considered as [8]:

$$Fe^{2+} \rightarrow Fe^{3+} + e^- \quad E^0 = -0.77 \text{ V} \tag{4}$$

$$2Fe^{2+} + \frac{1}{2}O_2(g) + H_2O(l) \rightarrow 2Fe^{3+} + 2OH^-$$

$$E_0 = -0.37 \text{ V} \tag{5}$$

Cathodic reactions [2]:

$$2H_2O + 2e^- \rightarrow H_2(g) + 2OH^- \quad E^0 = -0.83 \text{ V} \tag{6}$$

Additionally, when chloride is present and the anode potential is sufficiently high, the following reactions may take place in the EC cell [9]:

$$2Cl^- \rightarrow Cl_2 + 2e^- \quad E^0 = -1.36 \text{ V} \tag{7}$$

$$Cl_2 + H_2O \rightarrow HClO + H^+Cl^- \quad E^0 = -0.93 \text{ V} \tag{8}$$

$$HClO \rightarrow H^+ + OCl^- \tag{9}$$

The formation of active chlorine species (Cl_2, HClO, OCl⁻) enhances the performance of the EC reactor through oxidation reactions. The dissolution of the anode metal is based on Faraday's law:

$$m_{metal} = \frac{I \times t_s \times M}{z \times F} \tag{10}$$

where I is the applied current (A), t_s is the treatment time (s), M is the molar mass of the electrode material (M_{Al} = 26.982 g/mol, M_{Fe} = 55.845 g/mol), z is the valency of ions of the electrode material (z_{Al} = 3, z_{Fe} = 2), and F is Faraday's constant (96485 C/mol).

It has been found that the theoretical amount of anodic dissolution is exceeded in real EC applications. This phenomenon is referred to as superfaradaic efficiency, and the experimental values of anode metal dissolution have varied between 105% and 190% of the theoretically expected value [10-15]. This phenomenon is thought to be attributed to pitting corrosion, especially in the presence of chlorine ions [2].

Electrochemically generated metal cations will react spontaneously, forming various monomeric species such as $Al(OH)^{2+}$, $Al(OH)^{2+}$, $Al_2(OH)_2^{4+}$, and $Al(OH)^{4-}$ and polymeric species such as $Al_6(OH)_{15}^{3+}$, $Al_7(OH)_{17}^{4+}$, $Al_8(OH)_{20}^{4+}$, $Al_{13}O_4(OH)_{24}^{4+}$, and $Al_{13}(OH)_{34}^{5+}$, which finally transform into $Al(OH)_3$ according to complex precipitation kinetics [16]. Ferric ions generated electrochemically may form monomeric ions, ferric hydroxo complexes with OH^- ions, and polymeric species.

These species/ions are: $FeOH^{2+}$, $Fe(OH)_2^+$, $Fe_2(OH)_2^{4+}$, $Fe(OH)_4^-$, $Fe(H_2O)_5OH^{2+}$, $Fe(H_2O)_4(OH)_2^+$,

$Fe(H_2O)_8(OH)_2^{4+}$, and $Fe_2(H_2O)_6(OH)_4^{2+}$, which further react to form $Fe(OH)_3$ [17-19]. The formation of these complexes depends strongly on the pH of the solution. Above pH 9, $Al(OH)^{4-}$ and $Fe(OH)^{4-}$ are the dominant species [20].

Aluminum and iron hydrolysis products then destabilize pollutants present in the solution, allowing agglomeration and further separation from the solution by settling or flotation. Destabilization is achieved mainly by means of two distinct mechanisms, i.e. 1) charge neutralization of negatively charged colloids by cationic hydrolysis products; and 2) "sweep flocculation", where impurities are trapped and removed in the amorphous hydroxide precipitate produced. Several factors such as pH and coagulant dosage have an impact on the relative importance of charge neutralization and sweep flocculation. Microbubbles (H_2 and O_2) released at the electrode surfaces bring about electroflotation by adhering to agglomerates and carrying them to the water surface [21].

The most important factors influencing the efficiency of the EC process are the electrode materials used, applied current density, treatment time, and solution chemistry, including initial pH and the chemical composition of the aqueous solution being removed. The solution temperature, type of salt used to raise conductivity, presence of chloride, electrode gap, passivation of the anode, and water flow rate also have an impact on the removal efficiency and economic durability of a given EC application.

The advantages of EC over conventional coagulation (CC) include economic aspects (relatively low investment, maintenance, energy, and treatment costs), significantly lower volume of sludge produced, better sludge quality (lower water content, much larger and more stable flocs with better settlability), similar or slightly better efficiency, avoidance of chemical additions, ease of automation, simple equipment and compact size of EC systems (allowing decentralized treatment), greater functional pH range and pH neutralization effect, and the presence of electroflotation (EF) [5].

Evaluation Principles

The removal efficiencies (R%) presented in Chapters 3.1 - 3.7 have been calculated with the Equation (11):

$$R\% = \frac{c_0 - c_1}{c_0} \times 100$$

(11)

where c_0 and c_1 are pollutant concentrations before and after EC treatment, respectively. Hydraulic retention times (HRT, min) were calculated with Equation (12):

$$HRT = V/Q$$

(12)

where Q is the flow rate (l/min). Current densities (i, A/m^2) can be calculated with the Equation (13):

$$i = I/A_{eff}$$

(13)

where A_{eff} is the effective, submerged area of the anode (m^2). When no current density values were given by the authors, either the current or voltage value in optimum conditions is presented in Tables 1-7. The electrical energy consumption (EEC, kWh/m^3) presented in Chapters 3.1 - 3.7 has been calculated with the Equation (14):

$$EEC = \frac{U \times I \times t}{60 \times V}$$

(14)

where U is the applied voltage (V), t is the treatment time (min) and V is the volume of the treated water (dm^3). Operating costs (OC, EUR/m^3) have been calculated with the Equation (15):

$$OC = a \times EEC + b \times EMC \qquad (15)$$

where a and b are the current market prices of electricity (EUR/kWh) and electrode materials (EUR/kg), respectively, and EMC (kg/m³) is electrode material consumption. The market prices used in calculating economic value have varied slightly from one paper to another, depending on the country and year of publication (prices have been on the rise over time), but they have been within a very similar range. Derived from Equations (14) and (15), EEC and OC per kg of specific pollutant/parameter (kWh/kg$_x$ and EUR/kg$_x$) can also be readily calculated based on the removal percentage and initial/final concentrations of the pollutant. These values have been presented also by some of the authors. The pH values presented in Tables 1-7 represent the range in which the EC application performed the best (the highest value is marked in brackets), even though the process would have performed nearly as well with pH values outside of this range. If the natural pH of the aqueous solution was inside this range, it has been noted separately.

In a handful of papers, the author did not present economic values and current densities in optimum conditions, but they could be approximated by using the values (total submerged anode surface area, applied current and voltage or current density, volume of the wastewater treated, treatment time, initial concentration, removal efficiency, etc.) given in the article. It should be strongly underlined that these rather simple calculations were done only when the authors of the corresponding papers had not presented the numbers themselves but had clearly stated the values needed for the calculations, or when the values could be easily deducted from the publication in question. The approximation calculations were based on Equations (11) and (13)-(15). Current market prices were estimated to be approximately 0.10 - 0.11 EUR/kWh (in Finland in July 2011, including electrical energy, distribution of electricity, and taxes) and 1.6 - 1.7 EUR/kg for aluminum and 0.33 - 0.37 EUR/kg for iron. These values are also similar to those used in the reviewed papers. All currencies (usually United States dollar) given in the papers for OC values have been converted to euros (in Tables 1-7) using up-to-date exchange rates.

Additionally, in some articles all the optimum values were not clearly stated or no specific values were given. In such cases, the missing values have been estimated, if possible, and if the results were reasonable and in line with the text. Estimations were based on the figures, tables, and text presented. Also, in a few papers additional optimum parameters were taken into account, meaning e.g. that when a major drop in treatment time, current density, or EEC-value was found to correspond with only a slight reduction in removal efficiency, the lesser removal efficiency (and corresponding other) values were also considered optimal in economic terms. Whenever any of these actions have been performed, it has been marked accordingly in Tables 1-7.

Table 1: Recent applications of EC in the treatment of tannery, textile and colored wastewater

Water & wastewater types used	Genuine (G)/ Synthetic (S) water	Anode/ Cathode material	Reactor type	Volume treated [ml]	Optimum electrode gap [mm]	Optimum current density, treatment time and initial pH [A/m²], [min], []	Initial pollutant levels [mg/l]	Optimum removal efficiency [%]	Optimum EEC [kWh/m³]	Optimum EEC [kWh/kgx]	Optimum OC [€/m³]	Optimum OC [€/kgx]	Research group & Publication year
Tannery wastewater containing organic and inorganic pollutants	G	Fe	Batch	3000	50	22.4 20 (7 - 9)ᶜ	COD: 4100 - 6700 BOD: 630 - 975 Cr: 11.5 - 14.3 TSS: 600 - 955 TKN: 144 - 170 TDS: 13300 - 19700 O & G: 638 - 780 Color: 3800 - 6330 [Pt-Co]	COD: 95 BOD: 96 Cr: 100 TSS: 96 TKN: 62 TDS: 50 O & G: 99 Color: ~98ᵍ	0.13	n.d.	0.25	n.d.	Kongjao et al. 2008 [11]
Tannery wastewater	G	Fe	Batch	400	60	333 5/30ʰ 7.4ᵈ	COD: 3700 Sulfide: 440 Cr_total: 22 SS: 2690	COD: 46ᵏ/56ʰ Sulfide: 25/97ᵏ Cr_total: n.d./97 SS: n.d./70	~(3.13/ 15.63 - 16.49)ᵉ	1.8/~(9.0 - 9.5ʰ) kWh/kg_COD	n.d.	n.d.	Apaydin et al. 2009 [26]
Industrial textile wastewater	G	Al	Batch	2200	20	80 70 7ᶜ	COD: 1260 Turbidity: 1310 [NTU] TS: 1750	COD: 70 Turbidity: 90 TS: 50	n.d.	n.d.	n.d.	n.d.	Zodi et al. 2010 [28]
Wastewater containing dyes from textile industry (Direct Red 81, azo dye)	S	Al	Batch	500	15	18.75 60 5 - 9 [6]	Dye: 25 - 200 (optimum 50)	Dye: 98	n.d.	n.d.	n.d.	n.d.	Aoudj et al. 2010 [29]

Dye-containing wastewater		Electrode	Mode					pH	Concentration	Removal (%)					Reference
Dye-containing wastewater (Direct Red 23 & Reactive Blue 140)	S × 2	Fe	Continuous hydrogen gas collecting	4420	8	30/40	5/5	n.d.	Dye: 100/100	COD: 93/n.d. Color: 99/89 TS: 89/n.d.	0.69/1.42	n.d.	>0.12	n.d.	Phalakornkule et al. 2010 [30]
Blue reactive (Reactive Blue 140), red disperse (Disperse Red 1), mixed dyes and a real textile wastewater	S × 3 + G	Al Fe[b] Al-Fe	Batch	1800	8	30 - 40	5	(7 - 9.6)[c]	COD: 278 - 736 Color: n.d. TSS: 85 - 354 TDS: 1715 - 6106 TS: 1800 - 6460	[COD: ~90/ (55 - 79) Color: ~100/ (79 - 97) TSS: n.d./ (55 - 96) TDS: n.d./ (21 - 23) TS: n.d./ (26 - 28)][g]	~(0.3 - 1.0)[f]/ (0.42 - 1.62)[g]	n.d.	n.d.	n.d.	Phalakornkule et al. 2010 [31]
Crystal violet solution	S	Al/SSFe /SS[b]	Batch	500	11	28	5	5.8	COD: n.d. Dye: 50 - 200 (optimum 200)	COD: ~100[g] Dye: ~96[g]	0.4	n.d.	n.d.	n.d.	Durango-Usuga et al. 2010 [32]
Acid green dye 50	S	Al	Batch	n.d.	10	16.7	21	6.9 - 11 [9]	COD: n.d. Dye: 100 - 300 (optimum 100)	COD: 87 Color: 96[g]	~0.48[e]	3.82 kWh/kg_{COD} ~4.8 kWh/kg_{dye}[e]	n.d.	n.d.	El-Ashtoukhy & Amin 2010 [33]
Acid brown 14 (azo dye)	S	Al Al/Fe[b] Fe	Batch Batch (pilot scale)	500 9000	10 30	6.329 2.58	18 198	4 - 6.4 [6.4][e]	COD: n.d. Dye: 30 - 100 (optimum 50)	COD: 87 64 Color: 91 80	~(0.053 - 0.037)[e]	~(1.21 - 1.15)[e] kWh/kg_{COD} ~(1.16 - 0.92)[e] kWh/kg_{dye}	n.d.	n.d.	Parsa et al. 2011 [34]
Acid red 14 (azo dye)	S	Fe/Steel	Batch	250	10	100	4	4 - 10 [7]	Dye: 50	Color: 91	n.d.	n.d.	n.d.	n.d.	Aleboyeh et al. 2008 [35]
Orange II (azo) dye solution	S	Al	Continuous	3000	6	120	8.6[a] (0.35 l/min)	6.5	Dye: 10 - 50 (optimum 10)	Color: 95	n.d.	n.d.	n.d.	n.d.	Mollah et al. 2010 [36]
Reactive black 5 (azo dye)	S	Al Fe[b]	Batch	500	25	45.75	5	5 - 9 [5]	Dye: 40 - 200 (optimum 100)	Color: 99	~0.50[e]	4.96 kWh/kg_{dye}	n.d.	n.d.	Şengil & Özacar 2009 [12]
Indigo carmine (Acid Blue 74)	S	Mild steel/SS	Batch	1000	3	10.91/ 54.57[h]	180/ 15[h]	5 - 9 [8]	Dye: 25 - 100 (optimum 40)	Color: 99/91[h]	~0.68/0.22[e]	17/18[h] kWh/kg_{color}	n.d.	n.d.	Secula et al. 2011 [37]
Red dye solution mixture of 2-naphtoic acid & 2-naphtol	S	Al	Continuous	n.d.	10	312.5	5[a] (0.62 l/min)	6 - 9 [6]	COD: 2500 Dye: 25 - 200 (optimum 80 - 100)	COD: >80/~80[h] Color: 95/~91[h]	~0.30[e]/n.d.	3.2/1.6[h] kWh/kg_{dye} ~0.15[e]/n.d. kgCOD	n.d.	n.d.	Merzouk et al. 2009 [38]

[a]= HRT (hydraulic retention time) in EC systems with continuous mode of operation [min]; [b]= Observed as the best electrode configuration of those tested; [c]= The natural, unmodified pH value of the water or wastewater (found optimal); [d]= The natural, unmodified pH value of the water or wastewater (the effect of pH not researched); [e]= Approximation calculation based on values given in the article at issue; [g]= Optimum value estimated from the data in the article (precise value not given); [h] = Additional "optimum value" estimated from the data in the article; n.d. = Not determined.

OVERVIEW OF DIFFERENT TYPES OF WATER AND WASTEWATER RECENTLY TREATED BY ELECTROCOAGULATION

Chapters 3.1 - 3.7 present a summary of recent applications of EC with different types of water and wastewater divided into categories by topic. Removal efficiencies, economic values and essential operational parameters in optimum process conditions are presented in Tables 1-7 along with other specifications (i.e. electrode materials, genuineness/ artificiality and initial pollutant levels of the water, reactor type, volume of water treated, electrode gap) of the research in question. If multiple electrode materials were tested, the optimum values presented are for the electrode configuration found best (if any), which is noted in the corresponding columns in Tables 1-7.

Tannery, Textile and Colored Wastewater

The global tannery industry represents an important economic sector in many countries. The quantity of effluent generated is about 30 l for every kilogram of hide or skin processed and it contains high concentrations of organic pollutants and Cr (III), which could be oxidized to highly toxic and carcinogenic Cr (VI) [22,23].

Dye-containing wastewaters are a major environmental concern because of their unaesthetic nature and their ability to hinder the penetration of light into water, which is detrimental to living organisms in bodies of water [24,25]. Azo dyes are one of the most widely used synthetic dyes. They can be toxic and mutagenic to aquatic life and are molecularly stable, rendering them resistant to biological and even chemical degradation [24]. Table 1 presents a summary of recent applications of EC in the treatment of tannery, textile and colored wastewater.

A study has been conducted on the treatment of wastewater from a tannery plant using the EC technique. A bench-scale system with iron electrodes was employed. The wastewater had high initial pollution parameter levels (see Table 1). After optimization, the EC treatment was found very effective and produced clear water (see Figure 1). The natural pH of the wastewater (7.0 - 8.7) was found to be within the optimum range of values. The EEC and OC values were found to be low, 0.13 kWh/m^3 and 0.25 EUR/m^3, respectively. Parallel monopolar connections were found to be more suitable for the treatment process than monopolar serial and dipolar parallel connection modes [11]. Another investigation compared EC and EF (Electro-Fenton process - the addition of H$_2$O$_2$ to an EC process to bring about oxidation reactions) in treating genuine, highly polluted tannery wastewater [26]. It is worth mentioning that treatment of tannery wastewater by conventional biological methods is often inadequate for complete purification, especially of ammo- nia and tannins (low biodegradability due to polyphenolic compounds) [27].

Figure 1: Tannery wastewater (a) before and (b) after treatment by electroco-agulation. Adapted from [11].

Table 2: Recent applications of EC in the treatment of paper industry wastewater

Water & wastewater types used	Genuine (G)/ synthetic (S) water	Anode/ cathode material	Reactor type	Volume treated [ml]	Optimum electrode gap [mm]	Optimum current density, treatment time and initial pH [A/m²], [min], []	Initial pollutant levels [mg/l]	Optimum removal efficiency [%]	Optimum EEC [kWh/m³]	Optimum EEC [kWh/kgx]	Optimum OC [€/m³]	Optimum OC [€/kgx]	Research group & publication year
Black liquor wastewater (from pulp and paper industry)	G	Al^b / Fe	Batch	300	5	140 50 5 - 7 [7]	COD: 7960, Polyphenols: 3220, Color: n.d., TSS: 1160	COD: 98, Polyphenols: 92, Color: 99, TSS: n.d.	n.d.	n.d.	n.d.	n.d.	Zaied & Bellakhal 2009 [45]
Bleached kraft mill effluent (pre-treated by sedimentation & aeration)	G	Al^b / Fe	Batch	250	20	48 2/7.5^b 7.6^g	COD: 426, BOD: 26, Lignin: 13514, Phenol: 0.535	COD: ~46^e/75^b, BOD: ~34^b/70^b, Lignin: ~79^e/80^b, Phenol: ~96^e/98^b	~(0.12/ 0.46)^e	~-120/~350 kWh/kg_COD^g, ~-3.2/~9.5 kWh/kg_lignin^e	n.d.	n.d.	Uğurlu et al. 2008 [46]
Paper mill wastewater (gravitationally & biologically pre-treated)	G	Al^b,g / Fe	Batch	2500	20	150^g 90^g 7.7^d	COD: 285, DOC: 75, Turbidity: 35 [NTU], Lignin: n.d, As: 3.8	COD: 68, DOC: 46, Turbidity: 100 (Fe), Lignin: 50 (Fe: 75), As: 92	n.d.	n.d.	n.d.	n.d.	Zodi et al. 2011 [47]
Paper mill wastewater	G	Al, Al/Fe, Fe/Al, Fe	Batch	1500	10	700 30 (5 - 7)^c	COD: 1700, Color: n.d., Phenol: 34, BOD: 850, TOC: 910, TSS: 1060, TS: 9801	COD: 86^g (Fe), Color: 92^b (Al), Phenol: 96^g (Fe), BOD: n.d., TOC: n.d., TSS: n.d., TS: n.d.	n.d.	n.d.	n.d.	n.d.	Katal & Pahlavanzadeh 2011 [48]
Tissue paper wastewater	G	Al	Batch	300, 4500 (scale-up)	10, 10	10.8 - 16.2, 10.8 30, 60 (6.7 - 7)^d	Turbidity: 80 - 120 [NTU], COD: 2645, BOD: 1688, TSS: 205, TDS: 1200 - 1600	Turbidity: 92 - 97, COD: n.d., BOD: 50, BOD: 60, TSS: n.d., TDS: n.d., n.d.	0.62 - 1.17, 0.68	n.d.	n.d.	n.d.	Terrazas et al. 2010 [10]

^b = Observed as the best electrode configuration of those tested; ^c = The natural. unmodified pH value of the water or wastewater (found optimal); ^d = The natural. unmodified pH value of the water or wastewater (the effect of pH not researched); ^e = Approximation calculation based on values given in the article at issue; ^g = Optimum value estimated from the data in the article (precise value not given); ^h = Additional "optimum value" estimated from the data in the article; n.d. = Not determined.

Biological treatment of wastewater containing resistant and toxic compounds requires long processing times and large treatment areas and generates high amounts of low-density sludge.

The experiments were repeated three times and the experimental error was found to be around 3%. The results presented are for the EC process only. It was found that the EF process was 10% more efficient in removing pollutants while its energy consumption was 20% lower. However, the cost of adding hydrogen peroxide was not considered. It was concluded that both processes showed fast and efficient purification of tannery industry wastewater. Note that the latter number presented in Table 1 is a visual approximation of Figure 1 in the article, because the precise value was not given in the text [26].

Genuine dark-grey colored, turbid (initially 1310 NTU, nephelometric turbidity unit) textile wastewater was treated by a recirculated batch EC process using Al electrodes. The natural pH of the wastewater was 7, which was also the optimum value, making addition of chemicals unnecessary. Optimization of removal efficiency with response surface methodology (RSM) corresponding to the Box-Behnken experimental design was successfully performed. Statistical testing of the model obtained was conducted using Fisher's statistical test for analysis of variance (ANOVA). Percentages of COD (chemical oxygen demand), TS (total solids) and turbidity removed were taken as the system responses, while current density, initial pH, and treatment time were the input parameters [28].

A batch EC system with Al electrodes was proposed for decolorizing synthetic azo-dye-containing industrial wastewater. Direct Red 81-dye concentration was measured using ultraviolet-visible spectrophotometry (maximum absorbance at wavelength I_{max} = 522 nm). It should be noted that this procedure was also used to estimate dye concentrations in other works presented in Table 1 when synthetic dye solutions were used, employing relevant maximum intensities. In this study, however, four supporting electrolyte types were compared; of these, NaCl showed the best performance. This was suggested to be due to Cl⁻ anions destroying the passivation layer formed on the aluminum electrode, leading to a greater rate of anodic dissolution. A decolorization rate of 98% was reached in optimum conditions [29].

Two different types of synthetic dye effluents were prepared and treated with an EC apparatus working in a continuous upflow mode. This EC system also applied a hydrogen gas collecting system. The dyes used to prepare the wastewater were azo-based Direct Red 23 and Reactive Blue 140. In Table 1, the first number presents the wastewater containing Direct Red 23; the latter is the wastewater with Reactive Blue 140. Experimental and theoretical maximum hydrogen yields were compared, denoting 89% - 94% efficiency in the gas collecting system. The results showed that the energy yield of harvested hydrogen (converted to electricity for the EC process with an assumed efficiency of 50%) could reduce the EEC-value of the EC process by 13% and 8.5% for Direct Red 23 and Reactive Blue 140 solutions, respectively. It was also stated that the high-quality hydrogen collected could also have been saved for use as a reactant in industrial processes. Decolorization rates of 99% and 89% for a 5-min EC-run applying a current density of 30 A/m^2 and 40 A/m^2 were found to be the optimum conditions for the Direct Red 23 and Reactive Blue 140 solutions, respectively. EEC-values were found to be low and the OC of the EC system was calculated as being less than 0.12 EUR/m^3 [30].

A typical textile effluent may have fluctuating properties because it contains various types of dye molecules. Therefore, a study was set up to investigate the decolorization of two different dyes (Reactive Blue 140 and azo-based Disperse Red 1) and a mixed dye made of them. Real textile wastewater was also treated with the same batch-EC system. All three synthetic wastewaters had results very close to each other (both EEC-values and removal efficiencies, which showed complete decolorization), therefore their values are given as their approximate averages in Table 1, followed by the results for real wastewater. Three different electrode configurations were tested, of which iron electrodes proved to be superior. The same optimum parameter values that were used for the synthetic dyed wastewaters prior to it were also used for the treatment of real wastewater, thus these values might have not been optimal for it. Of the five EC tests run in these conditions, one seemed to have failed (it was not in line with the others) and was therefore ruled out [31].

A batch EC system was employed to treat a synthetic crystal violet (CV) solution. Na_2SO_4 was used as the support electrolyte in this work, which claimed that NaCl was a controversial choice because of its possibility to form organic chlorine by-products. In this study, a twolevel full factorial experimental design (2^k) was employed to evaluate decolorization of the CV wastewater by EC. The levels of four variables (initial pH, CV concentration, supporting electrolyte concentration, current density) were studied. Reduced empirical models for both Al and Fe anodes were proposed for CV removal with EC. The correlation percentages were 96% and 83% for Al and Fe anodes, respectively. Iron was found superior to aluminum in this application, and residual amounts of less than 1 mg/l of iron were detected after an optimal 5-min EC run, while color and COD were fully removed [32].

A comparison of EC and EO (electro-oxidation) in treating Acid Green dye 50-based synthetic wastewater was done. Both processes were carried out in novel cathodic H_2-gas-stirred batch reactors. This was proposed to reduce the capital and operating costs of the reactor by making mechanical stirring unnecessary, provided that the EC cell is designed properly. EC was found more efficient: in optimum conditions, practically complete decolorization and a COD reduction of 87% versus 68% were accompanied by clearly lower energy consumption than what was achieved with EO. However, it was concluded that further studies on different types of dyes are needed to confirm this [33].

EC was employed to remove Acid Brown 14 from an aqueous solution by bench-(500 ml) and pilot-scale (9000 ml) batch processes. This type of dye was chosen because of its high level of usage in the textile industry. Aluminum was found superior to iron as the anode material for this application. Rather low values of current density (6.329/2.58 A/m^2) were found optimal for the batchand pilot-scale systems, leading to very low EECvalues (0.053/0.037 kWh/m^3) while achieving substantial removal efficiencies for COD and color. The EC process was concluded to be promising in treating azodye-containing wastewater [34].

Optimization of Acid Red 14 azo dye removal by a batch-EC (Fe/ steel electrodes) process with RSM was performed. Evaluation was based on the simple and combined effects of three main independent parameters: current density, treatment time, and initial pH of the wastewater. The study showed clearly that RSM was suitable for optimizing the EC process operating conditions and maximizing dye removal (91.27% in 4 min, whereas 93.93% in 4.47 min was predicted by the model, see Figure 2). A high coefficient of determination (R^2 = 0.928) ensured satisfactory adjustment of the model derived from the experimental data [35].

A continuous EC system with a 450-ml electrolytic cell bearing aluminum electrodes was utilized in treating synthetic wastewater polluted with azo-based Orange II dye. In this experiment, it was found clearly evident that when operating in optimum conditions (near-neutral initial pH, 350 ml/min flow rate, 4 g/l added NaCl), the color of the dye solution had almost completely vanished. The EC apparatus was summarized to be simple to design and operate and is an inexpensive tool for treatment of dye-containing textile wastewater [36].

Treatment of other artificially colored wastewater by batch EC was investigated, with iron used as the electrode material due to its clear superiority over aluminum here. The dye used in the experiments was azo-based Reactive Black 5, which was chosen because of its extensive annual consumption rate. Complete decolorization of 100 mg/l initial dye concentration was achieved rapidly in 5 min, by adding of 3 g/l NaCl and applying a current density of 45.75 A/m^2. Optimal initial pH was found to cover a broad range of pH values [12].

An Indigo Carmine (Acid Blue 74)-based aqueous solution was treated with a batch EC system employing mild steel/SS electrodes. The gap between the vertically positioned electrodes was only 3 mm. Generation of green iron(II) hydroxide into the solution changed its color from dark blue to dark green and further to yellow brown. At the end of the EC process, stirring was halted and sedimentation and flotation occurred, and a removal efficiency of 96% was achieved in optimal conditions. The observed color change during the EC process is depicted in Figure 3. A wide range of initial pH values were found to be suitable for this application. It was also shown that controlling pH (preventing it from rising by acid addition) during the EC process was detrimental and limited the development of flocs, at least in this case. NaCl was proposed as a better choice than Na_2SO_4 as a supporting electrolyte [37].

A red dye solution of 2-naphtoic acid and 2-naphtol was treated with Al-EC in a continuous mode. The reactor consisted of two compartments, the first being the actual electrolytic cell and the second a settling compartment filled by overflow from the first one. The initial COD value of the solution was 2500 mg/l when the total concentration of the dye mixture was 100 mg$_{dye}$/l. Optimal results were achieved with a 14-min residence time (of which only 5 min in the EC compartment), resulting in 95% dye removal efficiency and a 3.2 kWh/m^3 EECvalue. The EEC-value could still be halved by increasing the conductivity of the water by adding NaCl, but this led to a slight reduction in removal efficiency. The high performance of EC in a continuous mode in removing this type of dye from wastewater was proven in this paper [38].

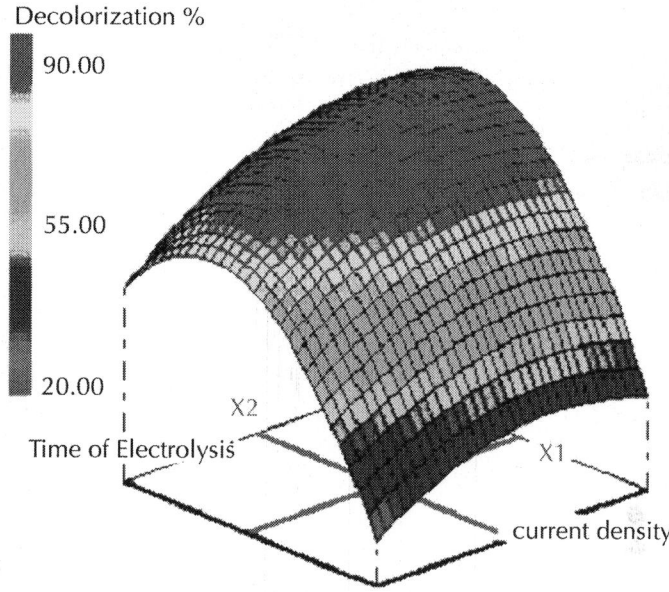

Figure 2: Three-dimensional contour plot obtained from the experimental data of color removal efficiency vs. current density (X1) and time of electrolysis (X2). Adapted from [35].

(a)

(b)

Figure 3: Evolution of the electrocoagulation process: (a) initial state; (b) 20 min, pH = 8.2; (c) 40 min, pH = 9; (d) 180 min of electrocoagulation and 5 min after stirring was turned off (initial concentration = 50 mg/l, $pH_{initial}$ = 7.1, current density = 10.91 A/m^2. Adapted from [37].

To summarize, EC treatment of tannery, textile, and colored waters has been under extensive development in recent years, with promising results. Wide research has been done in this field in addition to the studies presented in Table 1; e.g. those published in [8,25,39-43].

Pulp and Paper Industry Wastewater

The pulp and paper industry is one of the major waterintensive chemical process industries, contributing significantly to environmental pollution, e.g. in the form of black liquor. Blackish color, high amounts of organic load, suspended solids (mainly fibers), COD, and BOD (biological oxygen demand) are characteristic of effluents from pulp and paper industry. Arsenic may also be present. The strong color of the wastewater derives mainly from polymerization between lignin-degraded products and tannins. The drawbacks associated with conventional treatment techniques have made it necessary to develop more effective methods for treating this type of wastewater. Table 2presents a summary of recent applications of EC in the treatment of pulp and paper industry wastewater [44].

Genuine black liquor wastewater with a high concentration of pollutants and a high pH value of 12 was treated with a batch EC system. The main characteristics of the wastewater before treatment are presented in Table 2. Repeatability tests were performed under the same experimental conditions. A relative standard deviation (R.S.D) value of less than 3% was achieved, proving good repeatability of the EC process. Aluminum and iron were very close to each other in efficiency with different pollutants being removed. However, aluminum was slightly better overall and was chosen as the optimum. Also, Fe electrodes caused the water to turn green at first, and then to yellow and turbid due to Fe(II) and Fe(III) species generated, whereas the resulting effluent treated with Al was found very clear and stable [45].

The paper mill effluents from a modern bleached kraft mill were utilized in an EC study. The lignin concentration of the pretreated, brown effluent was extremely high, 13514 mg/l. All experiments were repeated twice, and the experimental error was approximately 4%. Aluminum was found superior to iron as the electrode material. The effect of initial pH was not studied in this work (the near-neutral original value was used). The high EECvalue of COD compared with that of lignin is related to its significantly lower initial concentration in the wastewater. The results of this study were found to suggest that EC is an effective alternative in paper mill effluent treatment [46].

Pre-treated wastewater from another paper mill with color, pH, and COD similar to those in the previous study was treated by EC in a batch mode. The wastewater was constantly circulated in the EC system with a peristaltic pump. A settling test was also conducted in a separate 46-cm-high glass column at the end of the test. All tests were done in triplicate. No clearly superior single electrode configuration could be found here, considering all aspects. Aluminium had somewhat better removal efficiency (except for lignin) than Fe, but the flocs it produced were clearly weaker in quality (SVI index values of 0.081 - 0.091 l/g and 0.207 - 0.310 l/g for Fe and Al, respectively) and more difficult to handle. The research group concluded that they will therefore use Fe in their consecutive studies. However, Al was chosen as the optimum here based on raw efficiency numbers. EC was proposed as a very effective tool for treating wastewater of the paper industry [47].

Highly polluted paper mill wastewater was treated by EC. No clearly superior single electrode configuration could be found of the four that were tested. Using Fe caused color reduction to be only 62%, and when using Al, COD and phenol reductions were 77% and 91%, respectively. Hybrid electrodes were the most constant in every parameter investigated, with results between those of Al and Fe. It should be pointed out that the optimum current density suggested (700 A/m²) was the highest one presented in Table 2 and significantly higher than in other studies. It was found that the water could be purified optimally directly without pH adjustment. After duplicating all tests, the experimental error was found to be below 4% [48].

Wastewater from the tissue paper industry was treated by EC with an aim to obtain water quality acceptable for reuse. Real wastewater with high pollution levels (see Table 2) was employed. A scale-up system of the Albatch EC process was also tried, with similar efficiencies and EEC-values. The results showed that a separation gap of 10 mm produces a faster build-up of sludge between electrodes. However, it yields more efficient removal of turbidity and lower energy consumption than larger gaps. EC was concluded to have proven to be an efficient method for removing turbidity from this type of wastewater, producing water of quality (8 NTU) suitable for reuse in the paper bleaching stage [10].

In addition to the studies presented in Table 2, at least [44,49,50] have also recently contributed to EC research in this particular sector. On the whole, treating paper industry wastewater by EC seems to be a feasible alternative and a subject of interest.

Oily Wastewater

Oily wastewaters with greatly varying compositions and very high pollutant levels are generated by various sources, such as petroleum refineries, discharge of bilge and ballast water, workshops, petrol stations, rolling mills, restaurants, edible oil and soap factories, as well as other general industrial sources. Oil-in-water can be found as free-floating oil, as an unstable oil/water emulsion, and also as a highly stable oil/water emulsion, which are all difficult to treat [7,51].

Table 3 presents a summary of recent applications of EC in the treatment of oily wastewater. In addition to the studies presented in Table 3, also [14,52-61] have recently studied oily wastewater purification by EC with promising results, which shows that there is a high interest in EC research in the field of oily wastewater purification.

A synthetic industrial oil-in-water emulsion was prepared and treated with EC. The original oil concentration of the emulsion was 5%, which corresponded to extremely high COD and turbidity values of 62300 mg/l and 29700 NTU, respectively. Nevertheless, very high RSM-optimized removal efficiencies of 90% for COD and 99% for turbidity were achieved in less than 22 min with ANOVA showing a high variance coefficient ($R^2 = 0.998$), ensuring satisfactory adjustment of the model with the experimental data [62].

Purification of oily wastewater resulting from washing the bilges of boats using EC working in a batch mode was studied. Iron was found to best Al as the electrode material. EC treatment with both materials was tested using monopolar (MP) and bipolar (BP) electrode configurations. The EECand OC-values for a MP configuration were found to be significantly lower than those of a BP configuration, whereas COD reductions were found rather similar, thus making Fe-MP the most feasible solution. The initial green color of the wastewater disappeared after EC treatment and the effluent became more transparent. A very low OC value was achieved alongside high removal efficiency. Note that in Table 3, C_{10}- C_{50} stands for C_{10}-C_{50}-hydrocarbons (indicators of raw oil-based hydrocarbons) [63].

Sunflower oil refinery wastewater with a natural pH value as low as 1.4 and COD as high as 15000 mg/l was treated in a batch EC reactor with aluminum electrodes. Na_2SO_4 and PAC (poly aluminum chloride) were added to the water to increase its conductivity and enhance coagulation. The addition of 0.5 mg/l PAC raised the removal efficiency of COD from 94.5% to 98.9% in optimum conditions (see Table 3). A significant initial pH adjustment to 5 - 7 was found to be required for optimal functionality. The treated effluent was very clear and its quality exceeded the local direct discharge standard and therefore EC was found very efficient in treating this type of wastewater [64].

An industrial waste emulsion containing fluorescent penetrating oil used in aeronautics was treated with EC using Al electrodes. The parameters shown in Table 3 are for the industrial-scale EC system only, as in this study a successful two-step scale-up (from a batch mode laboratory system to continuous pilotand industrialscale systems) of the EC process with very similar performance parameters was conducted. The values given for the aforementioned EC process in Table 3 were ones achieved after additional sand and carbon filters, which contributed approximately 5 - 10 additional percentage points to removal efficiency. In this study, the industrial-scale system used utilized an innovative partial re-circulation of sludge supernatant (see Figure 4 for a presentation of the EC process pilot prototype), which was found to enhance coagulation (increasing removal efficiency) without increasing the EEC-value and to diminish the sludge generation rate. This phenomenon was proposed to be due to introduction of a basal quantity of coagulant working as a coagulant initiator to the wastewater.

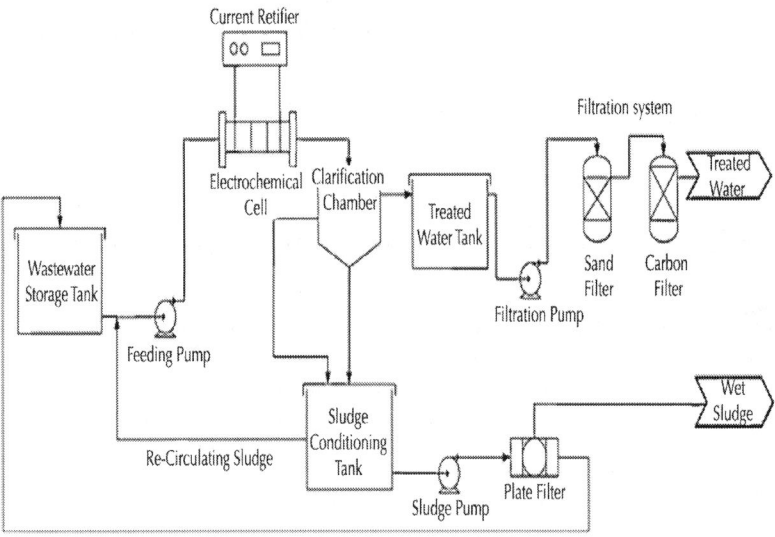

Figure 4: Scheme of the EC pilot prototype set up (see above). Adapted from [65].

Table 3: Recent applications of EC in the treatment of oily wastewater

Water and wastewater types used	Genuine (G)/ Synthetic (S) water	Anode/ cathode material	Reactor type	Volume treated [ml]	Optimum electrode gap [mm]	Optimum current density, treatment time and initial pH [A·m⁻²], [min], []	Initial pollutant levels [mg/l]	Optimum removal efficiency [%]	Optimum EEC [kWh·m⁻³]	Optimum EEC [kWh/kgₛ]	Optimum OC [€/m³]	Optimum OC [€/kgₓ]	Research group
Industrial oil-in-water emulsion	S	Al/SS	Batch	400	10	250 22 7	COD: 62300 Turbidity: 29 700 [NTU]	COD: 90 Turbidity: 99	3	n.d.	n.d.	n.d.	Tu & Moulai-Mostefa 2008 [62]
Oily bilgewater	G	Al Feᵇ	Batch	1700	15	1.5 [A] =~34ᵈ 60 4	COD: 3350-3450 TOC-468 BOD₅: 118-216 Turbidity: 2210 [NTU] C₁₀-C₃₉: 422-460 O&G: 800 TSS: 501-585 TS: 2160-2400	COD: 78 TOC: n.d BOD₅: 90-96 Turbidity: 98-99 C₁₀-C₃₉: 99-100 O&G: 95-96 TSS: 99-100 TS: 32-43	2.07-2.25	n.d.	0.032-0.033	n.d	Asselin et al. 2008 [63]
Vegetable oil refinery wastewater	G	Al	Batch Batch + PAC 0.5 mg/l	300	8	350 90 5-7 [?]	COD 15000	COD: 95-99	n.d	96 kWh/kgCOD - 42 kWh/kgCOD	~(0.39-1.2)ᵉ €/kgCOD	0.026-0.08	Un et al. 2009 [64]
Industrial wastewater containing a fluorescent penetrant oil (emulsion)	G	Al	Continuous + sand- and AC-filtration	23000	5	110 5ᵃ (3.2-7.5)ᵉ l/min) {6.5-7.5}ᶠ	COD: 1000-2500 Color: 350 [Pt-Co] Turbidity: 600 [NTU]	COD: 95 Color: 99 Turbidity: 99	~(0.5-0.7)ᵉ	n.d	0.18ᵍ	n.d	Meas et al. 2010 [65]
Olive mill wastewater	G (diluted to 20 %)	Al	Batch	100	28	250 15 (4-6)ᶜ	COD: 20000 Polyphenols: 260 Color: n.d	COD: 84 Polyphenols: 87 Color: 92	~5.26ᵉ	2.63 kWh/kgCOD	~0.54ᵉ €/kgCOD	0.27	Hanafi et al. 2010 [66]
Waste metal cutting fluids	G	Al Feᵇ	Batch	800	10	60 25 6-7 [7ᵃ]	COD: 17312 TOC: 3155 Turbidity: 15350 [NTU]	COD: 92 TOC: 82 Turbidity: 99	n.d.	n.d	0.348	0.0016 €/kgCOD 0.0101 €/kgTOC	Kobya et al. 2008 [67]
Contaminated groundwater (petroleum hydrocarbons)	G	(Al, Fe & SS in different combinations) SS/Feᵇ	Batch + aeration Continuous + aeration	200	30	180 (20/40 20/40ᵃ [0.01/ 0.005 l/min})ᵍ (6-9)ᶜ [?]	TPH: 64 TDS: 1178	(TPH: ~85/~91 ~81/ ~91)ᵍ TDS: n.d	n.d.	n.d	n.d	n.d.	Moussavi et al. 2011 [68]
Palm oil-based biodiesel wastewater	G	Al/ Graphite	Batch	1000	15	20 [V] = ~135-140ᵃ 25 6	COD: 30980 O&G: 6020 SS: 340 Methanol: 10667	COD: 55 O&G: 97 SS: 98 Methanol: n.d.	5.57	n.d	n.d	n.d	Chavalparit & Ongwandee 2009 [70]
Rose oil processing wastewater	G	Fe	Batch	400	65	80 20 6.4-7.1 [64]	COD: 9500 BOD: 4930 Turbidity: 750 [NTU] TS: 7690	COD: 80 BOD: n.d Turbidity: 81 TS: n.d	6.25	0.825	n.d	n.d	Avsar et al. 2007 [71]

ᵃ= HRT (hydraulic retention time) in EC systems with continuous mode of operation [min]; ᵇ= Observed as the best electrode configuration of those tested; ᶜ= The natural, unmodified pH value of the water or wastewater (found optimal); ᵈ = The natural, unmodified pH value of the water or wastewater (the effect of pH not researched); ᵉ = Approximation calculation based on values given in the article at issue; ᵍ= Optimum value estimated from the data in the article (precise value not given); n.d. = Not determined.

The removal efficiency was high while the OC value was found to be very low. Cost estimates for the process indicated an investment-return time for the EC system of only about 17 weeks (at a rate of 8 m³ wastewater generated weekly at the plant) compared with the plant's current policy of simply sending it out for disposal [65].

Genuine olive mill wastewater (OMW), diluted with water to a fifth of the original concentration, was treated with EC an system using aluminum electrodes. Even after diluting the OMW, initial pollution concentrations were extremely high, as can be seen in Table 3. All experiments were tripled, achieving accuracy better than 4%. The naturally occurring pH of OMW and a 2-g/l NaCl addition were found appropriate for achieving effective treatment, in which a significantly low EEC-value of 2.63 kWh/kg$_{COD}$ and very high removal efficiency (see Table 3) could be reached. The final pH of the OMW was nearly neutral. It was also found that the EC treatment reduced the toxicity of OMW for Bacillus cereus by 70%. The growth of bacteria was nearly similar to a standard medium in the treated OMW, whereas in untreated OMW, bacterial growth was impossible. Consequently, EC was considered a viable pre-treatment step prior to a biological process for treating of OMW [66].

Metal cutting fluids are widely used for cooling and lubrication in metal industries. Batch EC treatment was performed on an extremely highly polluted (see Table 3) white-colored waste metal cutting fluid (WMCF) obtained from one such company's production of automotive engines, transmissions, and stamping plants. Iron was found to be clearly more economical than aluminum in treating the WMCF, even though removal efficiencies were very similar and high for both electrode materials. Operating costs were found to be low and the natural pH of WMCF was found to be optimal, thus addition of chemicals to alter the initial pH was unnecessary. It was concluded that despite achieving high COD reductions, further treatment is nonetheless needed because of the residual COD values, which still exceeded the local discharge standards [67].

The efficacy of EC for treating petroleumcontamnated groundwater (from a site close to a local petroleum refinery) was evaluated and quantified as total petroleum hydrocarbons (TPH) removed. Al, Fe, and SS in different combinations were tested as electrode materials, of which a SS/Fe-combination proved to be the most suitable. The EC process was studied with systems working in both batch and continuous modes. The natural neutral pH of the groundwater was found to be optimal and increasing HRT was found to improve TPH removal in the continuous EC systems. Note that the removal efficiencies shown in Table 3 were achieved with aeration in the EC cell, which added approximately 22 percentage points to the removal efficiencies (see Figure 5). This was proposed to be due to aeration-transferred oxygen accelerating the oxidation of Fe^{2+} (dissolved from the anode) in solution to Fe^{3+}, which in turn resulted in formation of greater $Fe(OH)_3$ flocs, thus improving TPH removal by adsorption. EC was summarized to be a promising technique in eliminating TPH from groundwater [68].

Biodiesel production generates large amounts of alkaline and highly oily wastewater with low nitrogen and phosphorus content, rendering it unable to be purified by biological treatment [69]. Such wastewater (with frying oil waste and crude palm oil used as biodiesel feedstock) was treated with a batch EC system using aluminum as the anode and (rather rarely) graphite as the cathode material.

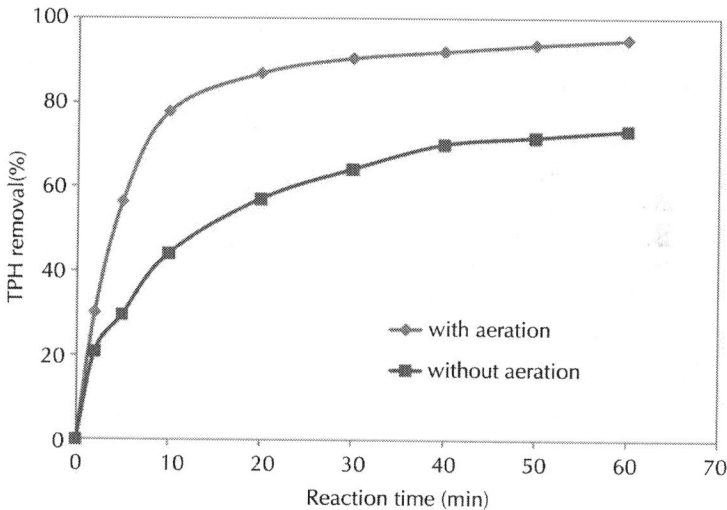

Figure 5: Effect of aeration on the removal of TPH in a batch-EC process (using Fe electrodes) as a function of reaction time. Adapted from [68].

Box-Behnken design-based RSM optimization was applied to evaluate the best process conditions. The RSM results matched experimental values very well; the predicted optimum values for initial pH, voltage, and reaction time were 6.06, 18.2 V, and 23.54 min, respectively, with removal efficiencies all within 0.7 percentage points of those obtained experimentally. Ultimately, EC was found suitable as a primary treatment of biodiesel wastewater, which, however, still requires further biological treatment [70].

Industrial wastewater generated in rose oil processing was purified by EC. In a previous study by the same research group, Fe was proven better as the electrode material than Al for this type of wastewater. This was due to clearly lower residual metal concentrations, which were less than 0.2 mg/l with Fe. In this paper, CC, EF, and two different Fenton processes were compared. It was concluded that the EC process was most suitable for this type of wastewater, while CC performed the worst. Initial pollutant concentrations in the wastewater were very high (see Table 3). Note that even though 6.4 was found to be the optimum initial pH value, an acidic natural pH value of 4 was used to achieve the efficiencies presented in Table 3, because the difference in removal efficiencies was negligible [71].

Food Industry Wastewater

Wastewaters from agro-industries come from a myriad of sources and their compositions vary greatly. On the whole they are, however, characterized by high COD and BOD due to their high level of organic content [72]. Table 4 presents a summary of recent applications of EC in the treatment of food industry wastewater.

The dairy industry is associated with generation of huge amounts of wastewater (approximately 0.2 l to 10 l of effluent per liter of processed milk) that is high in nu- trients and has highly varied pH [73,74].

Table 4: Recent applications of EC in the treatment of food industry wastewater

Water and wastewater types used	Genuine (G)/ Synthetic (S) water	Anode/ Cathode material	Reactor type	Volume treated [ml]	Optimum electrode gap [mm]	Optimum current density, treatment time and initial pH [A/m²], [min], []			Initial pollutant levels [mg/l]	Optimum removal efficiency [%]	Optimum EEC [kWh/m³]	Optimum EEC [kWh/kgx]	Optimum OC [€/m³]	Optimum OC [€/kgx]	Research group
Dairy wastewater	S	Fe	Batch	1500	10	270	50	~5	COD: 3900 Turbidity: 1744 [NTU] TS: 3090 TN: 113	COD: 70 Turbidity: 100 TS: 48 TN: 93	~(0.83 - 30.0)ᵃ	~2.76 kWh/kg_COD	0.051 - 1.30	n.d.	Kushwaha et al. 2010 [77]
Dairy wastewater	G	Fe	Batch	650	25	60	1	6 - 8 [7ᶜ]	COD: 18300 O & G: 4570 TSS: 10200	COD: 98 O & G: 99 TSS: n.d	~0.055ᵃ	0.003 kWh/kg_COD	n.d.	n.d.	Şengil & Özacar 2006 [73]
Potato chips manufacturing wastewater	G	Al / Fe	Batch	250	11	20 - 300	5 - 40	4 - 6	COD: 2200 - 2800 Turbidity: 260 - 610 [NTU] BOD: 1650 - 2150	COD: 60 Turbidity: 98 BOD: n.d.	4.9	n.d.	0.36 - 4.10	n.d.	Kobya et al. 2006 [78]
Almond industry wastewater	G	Al/Fe / Fe/Al	Batch Continuous (pre-industrial scale up)	700 ~54000ᵃ	10 15	50 45ᵍ	15 (6.67 l/min)	4 - 8 [5.7ᶜ]	COD: 5300 TOC: 1400 BOD₅: 1000 Color: 18000 [Pt-Co] Turbidity: 3200 [FTU] N_total: 240 P_total: 3 TSS: 3400	COD: 81/81 TOC: 74/79 BOD₅: 80/67 Color: 100/98 Turbidity 99/98 N_total: 75/85 P_total: 100/99 TSS: 100/99	n.d./2.8ᵍ	n.d.	n.d.	n.d.	Valero et al. 2011 [79]
Pasta and cookie processing wastewater	G	Al	Batch Batch + H₂O₂ (= EF)	1500	n.d	0.0182	60	3 - 8 [4]	COD: 7500 BOD₅: 3445 Color: 35 [Pt-Co] Turbidity: 1153 [NTU] TS: 2905 Fecal coliforms: MPN: 11000	COD: 80/90 BOD₅: n.d./96 Color: n.d./57 Turbidity: n.d./97 TS: n.d./95 Fecal coliforms: n.d./100	n.d.	n.d.	n.d.	n.d.	Roa-Morales et al. 2007 [80]
Poultry slaughterhouse wastewater	G	Al / Fe	Batch + polymer (LPM 9511, 10 mg/l)	1700	15	0.3 [A] = ~2ᵃ	60	(6.1 - 6.5)ᶜ	COD: 2700 - 3100 BOD: n.d. Turbidity: n.d. O & G: 720 - 950 TSS: n.d. TS: 1440 - 2380	COD: 80 - 84 BOD: 84 - 88 Turbidity: 86 - 94 O & G: 98 - 100 TSS: 85 - 93 TS: 58 - 70	4.07 - 4.31	n.d	0.49 - 0.51	n.d	Asselin et al. 2008 [81]
Poultry manure wastewater (UASB pretreated)	S	Al / Fe	Batch	400	62	150	20	4 - 6 [5]	COD: 4120 Color: 3390 [Hᵗ]	COD: 90 Color: 92	n.d.	2.6 kWh/kg_COD	n.d	n.d	Yeni-mezsoy et al. 2009 [15]
Egg processing wastewater	S - G	Al / Fe / SSᵇ	Batch Batch + coagulant (200 mg/l bentonite)	1000	70	15	30/ (35 - 26)	(4.5 - 5)ᶜ	COD: 8632 - 8983; 4068 - 4132 TSS: 1651 - 1953/936 - 1086 Turbidity: 933 - 1267; 1340 - 2060 [FTU]	COD: 97 (92.95) TSS: 97/(97.95) Turbidity: 99 (99, 99)	n.d	n.d	n.d	n.d	Xu et al. 2002 [82]
Baker's yeast wastewater	G	Al / Fe	Batch	800	2	70	50	5.5 - 6.5 [6.5] 6 - 8 [7]	COD: 2485 TOC: 1061 Turbidity: 2075 [NTU] TSS: 503	COD: 71/69 TOC: 53/52 Turbidity: 90/56 TSS: n.d	n.d.	n.d.	1.08/0.36 €/kg_COD	0.58/0.19 €/kg_COD	Kobya & Delipinar 2008 [83]
Tea factory wastewater	G	Steel	Batch	400	5	24 [V]	n.d	6ᶜ	COD: 293/607 BOD: 42/193 Color: 2604/9210 [Pt/Co]	COD: 91/97 BOD₅: 84/42 Color: 100/100	2.27/2.76	n.d	n.d	n.d	Maghanga et al. 2009 [84]

ᵃ= HRT (hydraulic retention time) in EC systems with continuous mode of operation [min]; ᵇ= Observed as the best electrode configuration of those tested; ᶜ= The natural, unmodified pH value of the water or wastewater (found optimal); [e]= Approximation calculation based on values given in the article at issue; [g]= Optimum value estimated from the data in the article (precise value not given); n.d. = Not determined.

In addition to those presented in Table 4, EC-treatment of dairy wastewater has been recently investigated by at least [75,76]. In a study on simulated dairy wastewater (SDW), a batch Fe-EC system was used coupled with a four-factor, fivelevel, full-factorial central composite design (CCD) based on RSM [77]. In optimum conditions, removal efficiency was high while economical values were low. Initial pollution levels of the SDW were high, with COD, TS, and turbidity being 3900 mg/l, 3090 mg/l and 1744 NTU, respectively. The authors suggested that, based on the results of TGA/DTA (thermal gravimetric analysis/ differential thermal analysis), the electrogenerated SDWsludge could be dried and used as fuel in boilers/incinerators or in fuel briquette production.

Real dairy effluent was treated by a Fe-EC system. Initial pollutant concentrations were very high; 18300 mg/l COD, 4570 mg/l O & G (oil and grease), and 10200 mg/l TSS (total suspended solids). Optimum treatment occurred very rapidly, as merely 1 min was found sufficient, while the EEC-value was extremely low (0.003 kWh/kg_{COD}) and removal efficiencies were very high (98% and 99% of COD and O & G, respectively). The equilibrium data obtained by the authors were found to fit very well into the Freundlich adsorption isotherm model ($R^2 = 0.99$) [73].

Potato chip manufacturing wastewater (PCW) was treated by batch-EC. Aluminum was found clearly superior to iron as the electrode material in this application. The removal efficiencies of COD and turbidity (having high initial values of 2200 - 2800 mg/l and 260 - 610 NTU, respectively) were high; however no clear single set of optimum process conditions was proposed. The natural pH of the PCW was 6.2 - 6.5, which was virtually optimal; therefore no pH adjustment was necessary. The results from kinetic studies showed that the kinetic data fit the second-order kinetic model very well ($R^2 > 0.96$) [78].

The results of batch and further continuous (removal efficiencies presented in Table 4 are in the form batch/ continuous) pre-industrial scale-up experiments showed that EC is an effective technology for treating wastewater from the almond industry, as the initially brown and murky wastewater turned colorless and clear as a result of EC treatment. Initial pollutant values were very high: e.g. 5300 - 6300 ppm COD and 3200 - 4000 FTU (formazin turbidity unit) turbidity. The results obtained for Al/Fe and Fe/Al electrodes were very similar, however the first was chosen over the latter mainly for practical reasons. The treated water was found to satisfy wastewater discharge legislation and the electrogenerated sludge was discovered to be neutral and non-toxic. This work also proved that EC can also be transferred (see Table 4, very similar results) to an automated industrial scale. No pH adjustment was required because the raw effluent had a pH value of 5.7, which was found optimal [79].

Wastewater from pasta and cookie processing was treated by an aerated batch -EC reactor after which the process was also investigated with added H_2O_2 (EF). The removal efficiencies presented in Table 4 are in the form EC/EF. The addition of H_2O_2 was found to raise the removal efficiency of COD from 80% to 90% with otherwise similar process conditions (values for other parameters with EC only were not given). The raw wastewater had very high initial pollutant values (see Table 4). A very interesting effect of total wastewater disinfection by EC was observed. The optimum initial pH range of EC was found to be very large in this application, being 3 - 8 [80].

Genuine red-orange-colored poultry slaughterhouse wastewater containing high amounts of organic matter (e.g. proteins, blood, fat) was treated by batch-EC using Al and Fe in monopolar (MP) and bipolar (BP) electrode configurations. The Fe-BP electrode configuration was found to be the most suitable here. Applying EC made the effluent more transparent and thus EC was found efficient for decolorization and clarification of poultry wastewater. After the optimum process conditions had been found, reproducibility tests were performed by repeating the EC test in triplicate. During these tests, 10 mg/l of a cationic polymer was added to the treated effluents. It was thus verified that the EC process is highly effective, economical, and repeatable (see Table 4) for treating this type of wastewater [81].

The performance of a batch-EC system was investigated on UASB (up-flow anaerobic sludge bed)-pretreated artificially generated poultry manure wastewater. High removal efficiencies of 90% COD and 92% color were achieved with 20-min EC-treatment. In this study, a toxicity test was also conducted using small fish, Lebistes reticulatus, which showed that the EC-treated water did not cause any of the fish to die or behave abnormally during the 48-h test (a local environmental requirement) or even after extended 120-h exposure to the water. Aluminum was found to outperform Fe as the electrode material, mainly because of the greatly lower color reduction efficiency of iron, while for COD removal, aluminum was found only slightly more effective. The initial COD concentration of the water was 4120 mg/l and it was dark brown in color (3390 Hazen units). The EECvalue was given only at the natural initial pH of 8.82, at which the results obtained were, however, very similar to those of pH 5. The results showed that almost 99.5% of the Al released precipitated in the form of EC sludge and the supernatant had an Al content of about 2.3 mg/l [15].

Simulated (SWS) and genuine egg processing industry wastewaters (EPW) with very high pollutant levels (see Table 4) were treated with batch-EC with multiple electrode compositions, of which SS was considered most suitable. The effect of adding 200 mg/l of bentonite coagulant was also studied using the real wastewater. The results presented in Table 4 are in the form SWS/(EPW, EPW + coagulant). The results obtained for both wastewaters showed very similar high removal efficiencies in a short treatment time. Additionally, EC yielded valuable by-products bearing high digestible protein and fat values. Addition of the coagulant further enhanced the EC process slightly. An economic analysis of EC indicated that this treatment is economically feasible and capital investments in equipment for a large-scale commercial egg processing plant could be recovered in 14 months [82].

Dark brown baker's yeast wastewater (BYW) with high pollution levels (see Table 4) was treated by batchEC. The results are presented in the form Al/Fe, because neither of the electrode materials used was proven to be universally superior to the other. Al was found to achieve slightly better removal efficiencies but at significantly higher operating costs. It was suggested that albeit the EC process could be adapted effectively for treatment of BYW, the effluent still contained a large amount of COD, which needed to be further treated by a secondary process [83].

Four wastewater samples (SP$_1$-SP$_4$) were taken from different points in constructed wetlands following the outlet of a tea factory, and their treatment by batch-EC with steel electrodes was tested. The results presented in Table 4 are those of waters SP$_1$/SP$_4$. The results of waters SP$_2$ and SP$_3$ were found rather similar. The waters were highly colored before the EC-treatment (2004/9210 mg/l Pt-Co), however 100% color removal in all the samples, SP_1-SP_4, was recorded. No NaCl was added to the wastewater samples even though their conductivities were low (134 - 317 µS/cm). Thus, it can be assumed that the energy consumption values could have been lower than the ones obtained. Diluting the waters prior to EC was found to be detrimental in terms of the EECvalue [84].

Other Types of Industrial Wastewater

Table 5 presents a summary of recent applications of EC in the treatment of other types of industrial wastewater.

Treatment of highly complex and highly polluted industrial wastewater (a mixture of wastewaters from 144 different factories received at a wastewater treatment plant) by batch-EC was studied. Three different electrode combinations were studied. Using aluminum and iron anodes simultaneously (corresponding cathodes were made of the same material as the anode) was found to outperform (better removal efficiencies, less sludge produced) the use of either metal alone as the anode material, combining the advantages of both. Complete disinfection of the wastewater was achieved. A short EC treatment time of 30 min resulted in almost similar (removal efficiencies only a few percentage points less) performance as a 60-min treatment time, for which the results are given in Table 5 [85].

Electroplating, metal finishing, and mining are industrial process in which large volumes of hazardous wastes containing heavy metals and free and metal cyanides are generated. Generally, cyanide removal from wastewater is carried out by chlorination, requiring high operating costs [86]. EC-treatment studies of separate cadmium and nickel electroplating rinse wastewaters (results in this order in Table 5) also containing high amounts of cyanide were carried out using iron electrodes in a batch mode. The raw pH values of both the wastewaters were found optimal. In optimum conditions (see Table 5), complete metal and cyanide removal was observed for both wastewaters, with OC-values of 1.05 EUR/m^3 (cadmium-laden wastewater) and 2.45 EUR/m^3(nickel-laden wastewater) [87].

Removal of Ni, Cu, and Cr from very heavily polluted industrial galvanic wastewater was carried out by batchEC. The raw pH of the wastewater was 1.5, which is strongly acidic, and adjustment to 5 was found necessary. The wastewater had very high conductivity (41 mS/cm) and its metal content was extremely high; around 2 g/l Ni, 2.5 g/l Cu, and 0.7 g/l Cr (70% present as Cr(VI)). The optimum electrode configuration of the EC system consisted of two separate anode-cathode-pairs used simultaneously instead of a single one made of Al or Fe. This novel EC process was found very efficient (see Table 5) in removing metals from galvanic wastewater. It was concluded that EC could be a good alternative or an after-treatment (the varying composition of such wastewaters may limit the feasibility of EC as a primary method) to conventional methods in this application [88].

Another study of Ni, Cu, and Cr removal from real industrial metal plating wastewater by batch-EC was conducted. The natural initial pH of the wastewater was 3 and it was chosen as "optimal" (optimal values presented in Table 5 given for this pH value) to avoid major addition of chemicals for pH adjustment, even though higher initial pH values (7 - 9) resulted in higher metal removal efficiencies, which were also obtained significantly faster. A Fe/Al electrode configuration was found most suitable of the four combinations of Al and Fe tested, however Fe/Fe was found to be nearly as efficient. In the end, it was concluded that EC in optimum process conditions could effectively reduce metal ions in the metal plating wastewater to a very low level, yielding 100% removal efficiencies for Ni, Cu, and Cr in 20 min [89].

Complexing agent and heavy metal (Zn, Ni) removal from a genuine, highly polluted metal plating effluent with SS-EC in a batch mode was studied.

Table 5: Recent applications of EC in the treatment of other types of industrial wastewater

Water and wastewater types used	Genuine (G)/ Synthetic (S) water	Anode/ cathode material	Reactor type	Volume treated [ml]	Optimum electrode gap [mm]	Optimum current density, treatment time and initial pH [A/m²], [min], []			Initial pollutant levels [mg/l]	Optimum removal efficiency [%]	Optimum EEC [kWh/m³]	Optimum EEC [kWh/kgₐ]	Optimum OC [€/m³]	Optimum OC [€/kgₐ]	Research group
Highly complex industrial wastewater (a mixture of wastewaters of 144 different factories)	G	Al Fe (Al + Fe)ᵇ	Batch	4000ᶠ	20	45,45	60	6 - 8 [8ᶜ]	COD: 2000 - 2500, BOD: 900 - 1050, Turbidity: 1400 - 1800[NTU], Color: 2500 - 4750 [Pt-Co], Fecal coliforms: MPN: 111000, TS: 5360	COD: 69, BOD: 71, Turbidity: 80, Color: 83, Fecal coliforms: ~99, TS: n.d.	n.d.	n.d.	n.d.	n.d.	Linares-Hernández et al. 2009 [85]
Electroplating rinse wastewaters containing cadmium, nickel and cyanide	G + 2	Fe	Batch	650	10	30/60	30/80	(7.6 - 10.6)ᵃ/ (8 - 10)ᶜ [10]	Cd: 102/-, Ni: ~175, Cyanide: 120/261, COD: 180/220, TSS: 175/185	Cd: 99/-, Ni: ~99, Cyanide: 100/100, COD: n.d/n.d, TSS: n.d/m.d.	6.13/11.94	n.d.	1.05/2.45	n.d.	Kobya et al. 2010 [87]
Galvanic wastewater	G	Al Fe (Al +Fe)ᵇ	Batch	1200	20	(1.0 [A] + 0.05 [A])ᵍ	180	5	Ni: 2000, Cu: 2500, Cr: 700	(Ni: 95, Cu: 100, Cr: 95)ᵍ	4.3ᵍ	n.d.	n.d.	n.d.	Heidmann & Calmano 2010 [88]
Metal plating wastewater	G	Al/Al Al/Fe Fe/Fe Fe/Alᵇ	Batch	650	10	100	20	7 - 9 [9]	Ni: 394, Cu: 45, Cr: 44.5	Ni: 100, Cu: 100, Cr: 100	10.07	n.d.	n.d.	n.d.	Akbal & Camcı 2011 [89]
Complexed wastewater from metal plating industry	G	SS	Batch	1800ᶠ	3	90	180	6 - 8 [6ᶜ]	TOC: 170 - 173, Zn: 217 - 236, Ni: 248 - 282	TOC: 66, Zn: 100, Ni: 100	n.d	160 kWh/kgₜₒc 70 kWh/kgₙᵢ	n.d.	n.d.	Kabdaşlı et al. 2009 [90]
Automotive assembly plant rinse waters from zinc phosphate coating	G	Alᵇ⁴ Fe	Batch Continuous	850 3500	11 20	60 60	25 (0.4/0.1 l/min)ᵍ [5]	3 - 6	COD: 150, TOC: 20, Turbidity: 80 [NTU], Zn²⁺: 40, Phosphate: 120, O & G: 10, TSS: 240	(COD: 77, TOC: 65, Turbidity: 98, Zn²⁺: 97, Phosphate: 99, O & G: 100, TSS: 97)ᵍ	~4.8ᵍ n.d.	n.d. n.d.	n.d. 2.98/6.74	n.d. n.d.	Kobya et al. 2010 [91]
Chemical mechanical polishing (CMP) wastewater from semiconductor fabrication	G	Fe/Al	Batch	500	20	20 [V]	20	(8 - 9)ᵈ	COD: 400 - 600, Turbidity: 200 - 300 [NTU], TOC: 3 - 5, TSS: 4000 - 5000	COD: 85, Turbidity: n.d, TOC: n.d, TSS: n.d	0.64	n.d.	n.d.	n.d.	Wang & Chou 2009 [93]
Carwash wastewater	G	Fe + BDD	Batch (EC) + Batch (EO)	300	15	20	6	6 - 9 [6.4ᶜ]	COD: 572, Surfactants: 95.5, BOD: 178	COD: 75/97, Surfactants: 100/100, BOD: n.d	0.14/12.0	n.d.	n.d.	n.d.	Panizza & Cerisola 2010 [94]
Alcohol distillery wastewater	G	Fe	Batch	1500	20	44.65	120	8	COD: 15600, BOD: 7200, TS: 34100, TDS: 2290	COD: 51, Color: 95, TS: n.d, TDS: n.d.	n.d	n.d.	n.d.	n.d	Kumar et al. 2009 [95]
Fermentation industry molasses process water (biologically pre-treated)	G	Al/SSᵇ Fe/SS	Batch	6250	6	137	57	7.5ᵈ	COD: 4500, Color: n.d, TDS: 1600	COD: 70, Color: 95, TDS: n.d.	n.d.	n.d.	n.d.	n.d.	Ryan et al. 2008 [96]
Coking wastewater	G	Fe/Ti	Continuous	2000	10	300	80ᵃ (0.025 l/min)	8 - 11 [8]	Color: n.d, COD: 91 - 111, BOD₅: 18 - 28, NH₃-N: 4.9 - 5.8	Color: 91, COD: n.d, BOD₅: n.d, NH₃-N: n.d.	n.d.	n.d.	n.d.	n.d.	Zhang et al. 2011 [97]
		Al	Batch	500	10	166.67	60	2 - 6 [2]	Sb_total: 10.4 - 28.6, As_total: 0.010 - 0.025, SBX: 0.376 - 0.434, Cu²⁺: 160 - 280	Sb_total: 97 - 98, As_total: 100, SBX: 71 - 77, Cu²⁺: n.d	n.d.	n.d.	n.d.		n

Laundry wastewater	G	Al	Batch	1500	15	1.32 [A]	45	(6 - 8)^c	COD: 4155 Turbidity: 245 [NTU] P: 27.6 Detergent: 463 Color: 1430 [n.d.] SS: 987 Pb: 4.35 Zn: 3.2	COD: 93 Turbidity: 96, P: 97 Detergent: 94 Color: 90 SS: n.d. Pb: n.d. Zn: n.d.	~19.8^e	n.d	n.d.	n.d	Janpoor et al. 2011 [99]
Paint manufacturing wastewater	G	Al^b Fe	Batch	800	10	35	15	4 - 8 [7]^c	COD: 19700 TOC: n.d SS: 1100 Pb: 1.44 Fe$_{tot}$: 4.82	COD: 94 TOC: 80 SS: n.d Pb: n.d Fe$_{tot}$: n.d.	n.d	n.d.	0.129	n.d	Akyol 2012 [101]
Marble processing wastewaters	G	Al^b Fe	Batch	250	5	15	0.5^b/2	(8 - 9)^e [9]	Turbidity: 2640 [FTU] SS: 5178 TDS: 0.21 O & G: 20	Turbidity: 99-100 SS: 99^b-100 TDS: n.d.-n.d O & G: n.d.-n.d.	-0.0568^e/-0.227^e	-0.0091^e/0.0363 kWh/kg$_{ss}$	-0.0155^e/0.062	-0.0025^e/0.01	Solak et al. 2009 [102]
Industrial aqueous solution containing polyvinyl alcohol (PVA)	S	Al/Al Al/Fe Fe/Al^b Fe/Fe	Batch	500	20	10 [V]	120	6.5^d	PVA: 100	PVA: 77	~1.57^e	15.68 kWh/kg$_{PVA}$	n.d	n.d	Chou et al. 2010 [103]
Industrial aqueous solution containing salisylic acid (SA)	S	Al	Batch	500	20	12	60^e	n.d.	SA: 100	SA: 87^e	n.d.	0.00037 kWh/kg$_{SA}$	n.d	n.d	Chou et al 2011 [104]
Water distribution system water	G	Fe	Batch	1500	20	12 [V]	60	10	Calcium hardness: 138 [mg/l CaCO₃] Total hardness: 300 [mg/l CaCO₃] Turbidity: 3 [NTU]	Calcium hardness: 98 Total hardness: 97 Turbidity: n.d.	n.d	n.d.	n.d.	n.d.	Malakootian et al 2010 [105]
Municipal wastewater	G	SS	Batch	1500^f	30	0.8 [A] = ~182^e	5	7^g	BOD$_{particular}$: 51 - 84 Turbidity: 49 - 53 [NTU] TSS: 126 - 160	BOD$_{particular}$: 99 Turbidity: 93 TSS: 95	n.d.	n.d.	n.d.	n.d	Bukhari 2008 [106]
Underground water used as feed water of a reverse osmosis (RO) desalination plant	G	SS + sand filter (2 L) & Birm iron filter	Batch	5000^f	n.d	1^b/1.75 [A] = ~(10 - 16/18 - 28)^e	6	7^g	Turbidity: 150 [NTU] TSS: 300 TDS: 1800	Turbidity: 92^b/98 TSS: 94^b/99 TDS: n.d./n.d	n.d.	n.d.	n.d.	n.d.	Sadeddin et al 2011 [108]
Mineral treatment wastewater containing ultrafine quartz suspensions	S	Al/SS	Batch	250	21	23.9^g	10	9	Turbidity: n.d	Turbidity: 90^g	1.87^g	n.d	n.d	n.d	Kılıç et al 2009 [109]
Bentonite suspensions (representing clay-polluted waters)	S	Fe	Batch	100^d	30	4	40	(3/12)	Turbidity: 18 - 24 [NTU]	Turbidity: 85-80	n.d	n.d.	n.d	n.d.	Ghernaout et al 2008 [110]

^{a}= HRT (hydraulic retention time) in EC systems with continuous mode of operation [min]; ^{b}= Observed as the best electrode configuration of those tested;^{c}= The natural, unmodified pH value of the water or wastewater (found optimal); ^{d}= The natural, unmodified pH value of the water or wastewater (the effect of pH not researched); ^{e}= Approximation calculation based on values given in the article at issue; ^{f}= Reactor volume (sample volume not mentioned); ^{g}= Optimum value estimated from the data in the article (precise value not given);

h= Additional "optimum value" estimated from the data in the article; n.d. = Not deter-mined.

The wastewater consisted of carrier, brightener, and metal chlorides strongly bound to the organic complexing agent (which comprised 90 % of TOC, total organic carbon). In optimum conditions (see Table 5), complete Zn and Ni removal was achieved and 66% of TOC was also removed. The natural pH of the wastewater was found to be optimal. NaCl additions were also found unnecessary due to the high electrolyte (1.5 - 1.7 g/l chloride) concentration in the wastewater. EC was concluded to be a promising treatment method for complexed metal removal from wastewater originating in the metal plating industry [90].

Zinc phosphate rinse water (ZPO) from an automotive assembly plant was treated by EC. Use of Al and Fe in batch and continuous modes was studied. The removal efficiencies of batch and continuous modes were very similar, thus the values presented in Table 5 are for the batch -Al process only. The removal efficiencies of Al and Fe were also very close to each other and no superior electrode material was suggested in the paper. However, Al was chosen as the optimum here because of its slightly better removal efficiencies and economic values. Al was also found to perform reasonably well in a significantly wider pH range than Fe. Furthermore, because the optimum initial pH for Fe would have been around 3, acid addition would have been necessary (the initial pH of ZPO was 3.8), whereas Al performed optimally with unmodified pH. In optimum process conditions, EC treatment was able to achieve high removal efficiencies with the pollutant parameters studied (see Table 5) [91].

Treatment of genuine chemical mechanical polishing (CMP) wastewater generated in the semiconductor fabrication industry by a batch -EC system was studied and found applicable. In another study on EC -treatment of CMP wastewater by the same research group, it was observed that Fe/Al is the most suitable electrode configuration (out of four different configurations tested) for such wastewater [92]. The CMP wastewater was highly alkaline and turbid (200 - 300 NTU), having a milky appearance while its mean particle size was as minuscule as 85 - 95 nm. Under optimum conditions, the EC process was found to remove 85% of COD in 20 min with a low EEC -value of 0.64 kWh/m^3. The very fast COD removal by the EC process was considered a great advantage of EC. The kinetic data obtained matched the pseudo first-order kinetic model well (R^2 = 0.97) [93].

Real carwash wastewater was treated by a combined EC/EO batch process. After Fe-EC, a 90-min EO-step at 100 A/m² current density was conducted using a boron-doped diamond (BBD) anode. The removal efficiencies and EEC-values presented in Table 5 are in the form EC/EC + EO total; other values are given for the ECstep only. After the EC-step in optimum conditions, 75% of COD was removed rapidly with low energy consumption (0.14 kWh/m³). Total surfactant removal was also noted after the EC -step. No pH alterations were found necessary. After the additional EO-step, 97% of COD was removed with a total EEC-value of 12.0 kWh/m³. In an earlier study by the same authors, only EO was used to treat the same wastewater, achieving complete COD removal, but at a cost of enormous energy consumption of 375 kWh/m³. Thus, adding EC prior to EO had lowered the EEC -value significantly while achieving similar removal efficiency [94].

High-strength (see Table 5) dark-black-colored biodigester effluent (BDE) from an alcohol distillery was treated by a batch -EC system using iron electrodes, employing RSM to optimize the process. The created model had a high R^2 value of 0.8547 and in optimum process conditions, 51% of COD and 95% of color were removed. It was also proposed that the EC-sludge of BDE could be used in making blended fuel briquettes along with other organic fuels, as its heating value was found to be 5.3 MJ/kg [95].

Variable and highly colored and polluted molasses process water (MPW) from the discharge outlet of an anaerobic/aerobic effluent treatment facility attached to a large industrial fermentation plant was treated by a batch EC system. In this study, CC using $FeCl_3$ and aluminum sulfate was compared with EC, which was tested with Al/SS (found superior) and Fe/SS electrode combinations. Both CC processes lowered the pH of MPW to strongly acidic values of 2.4 and 3.8, respectively (the natural pH of MPW was 7.5). However, EC raised the pH to mildly alkaline 8.6 - 8.8, so it was concluded by the authors that this makes EC a significantly better option. Also, at a major industrial plant with a 1000 m³/d output, 6 t of coagulant chemical would be needed, whereas only 300 kg of electrode material would produce similar results. Removal efficiencies were found to be in the same range with CC and EC. The EC process was not optimized properly at all, as the purpose of the work was solely to compare CC and EC. However, the results were still promising (see Table 5) and the reactor used was noticeably larger than in most other studies presented in Tables 1-7 [96].

A study on decolorization of coking wastewater containing inorganic pollutants and organic contaminants by a continuous EC process using Fe/Ti electrodes was conducted. The results showed great potential in EC-based decolorization (91% color removal efficiency) of coking wastewater with only a slight initial pH alteration from the natural 7 to 8 needed. Adding NaCl to the water showed a substantial increase in removal efficiency, possibly due to electrogeneration of Cl_2, a strong oxidant [97].

An EC technique with aluminum electrodes was used in a batch mode to remove toxic and carcinogenic antimony from antimony mine flotation wastewater. The antimony concentration of the water being treated was 10 - 30 mg/l. The water also contained As (10 - 25 µg/l), SBX (sodium butyl xanthate, 380 - 430 µg/l), and substantial amounts of cations; thus no NaCl was added to increase conductivity. The EC process performed almost as well in the initial pH range of 6 - 10 as in the optimal range of 2 - 6, indicating a wide scale of applicability and the redundancy of pH adjustment (the pH value of raw water was near 7). In optimum conditions (60 min of electrolysis at 166.67 A/m² current density), complete As removal and nearly complete antimony removal efficiencies were achieved along with 71% - 77% removal efficiency for SBX, indicating that EC is a promising technology for removing antimony from industrial wastewater [98].

Laundry-based wastewater accounts for approximately 10 % of municipal sewer discharges; thus the efficiency of a batch electrochemical system using aluminum electrodes in treating real laundry wastewater (see Table 5 for composition) was investigated [99]. All experiments were repeated twice and the experimental error was below 3%; average data are reported. The unaltered, nearneutral initial pH value was found optimal. Removal efficiencies in optimal conditions were high, being 90% - 97%. Therefore, it was concluded that when compared with other treatment processes, EC is more effective in treating laundry wastewater. In another study (not presented in Table 5) on EC treatment of (artificial) laundry wastewater, 62% COD removal efficiency was reached [100]. In this study, the application of ultrasound was studied and found to clearly enhance the EC process.

The treatability of paint manufacturing wastewater (PMW) by EC in a batch mode was investigated and found economic and feasible. The performance of Al electrodes was found to be better than that of

Fe electrodes in terms of removal efficiency and OC. Initial pollutant concentrations in PMW were very high (COD 19700 mg/l, BOD 2800 mg/l, SS 1100 mg/l), however in optimal process conditions (fast 15-min treatment at a low current density of 35 A/m^2), very high removals of 94% for COD and 89% for TOC were achieved. Absorbance decreased substantially as a result of the EC treatment, indicating a significant change in the color of the water. No pH alteration of the PMW was found necessary [101].

Removal of suspended solids and turbidity from marble processing wastewater by EC was studied using a batch laboratory-scale (250 ml solution) reactor. Both Al and Fe were tested as electrode materials. When iron was used as the electrode material, removal efficiencies were found to be only slightly lower than those of Al, but the OC values were significantly higher. Therefore, Al (monopolar parallel connection) was chosen as the better option. The initial concentrations of the wastewater were very high (turbidity 2640 FTU, TSS 5178 mg/l). EC treatment neutralized the wastewater, slightly lowering its pH value towards 7 from the initial optimal value of 9 (naturally 8.23). Complete removal of TSS and turbidity was achieved rapidly within 2 min and even only 0.5-min EC treatment was able to provide 99% removal efficiencies for TSS and turbidity. Therefore, the EEC and OCvalues were very low (see Table 5) and the EC process was concluded to be highly effective in this application [102].

Polyvinyl alcohol (PVA) is a well-known water-solu ble polymer that is hazardous and barely biodegradable. PVA is found in wastewaters of a wide range of industries; thus the feasibility of batch-EC in removing PVA from a synthetic (100 mg/l) aqueous solution was investigated. Of the four different electrode combinations tested, Fe/Al was found clearly the most efficient. The effect of altering the initial pH of the PVA solution was not studied. The experimental results showed that the kinetics of PVA removal by EC could be described with a pseudo-second-order model ($R^2 = 0.99$). In optimum conditions, 77% of the initial PVA was removed [103].

Salicylic acid (SA) is widely used in the pharmaceutical and cosmetic industries and it potentially has adverse health effects in animals and humans. EC -removal of SA (100 mg/l) from a synthetic industrial aqueous solution using aluminum electrodes in a batch mode was investigated and suggested to be promising. The effect of the initial pH of the SA solution was not studied and no base pH value was mentioned. Solution temperature was found to slightly affect removal efficiencies (up to about 9 percentage points in otherwise similar process conditions); 298 K was found optimal. In optimum conditions (applying a low current density of 12 A/m^2), 87% SA removal efficiency was documented, also providing an extremely low EEC -value. According to the kinetic data obtained, a pseudo-second-order kinetic model described SA removal best (R^2 = 0.98) [104].

The performance of Fe-EC in a batch mode in removing hardness from drinking water was evaluated. The water distribution system water used in this study had a pH value of 8.35 and total and calcium hardness values of 300 mg/l CaCO$_3$ and 138 mg/l CaCO$_3$, respectively. In optimum conditions, 98% of the former and 97% of the latter were removed, thus it was shown that ions responsible for water hardness could be removed by EC [105].

Raw municipal wastewater was electrocoagulated in a batch mode using SS-electrodes [106]. The EC-treatment was found effective and rapid, as it took only 5 min to achieve 99%, 93%, and 95% removal efficiencies for BOD$_{particulate}$, turbidity, and TSS, respectively. EC test runs were conducted using only raw wastewater with a pH of 7. In another study (not presented in Table 5) on actual municipal wastewater, it was concluded that by using very low currents (10 A/m^2), EC can reduce phosphorus and pollution associated with colloids, helping to diminish the organic load of the effluent [107].

Underground water (containing colloidal particles which cause membrane fouling in reverse osmosis, RO) used as the feed water of a RO desalination plant was treated with a batch EC-system using electrodes made of SS. A sand filter (2 l) also containing a Birm (a solid similar to active carbon, used for iron removal) filter was added to the process line after the EC unit to remove coagulated matter. The removal efficiencies given in Table 5 were achieved after the whole process. Before EC, the water had turbidity and TSS values of 150 NTU and 300 mg/l, respectively, and its pH was 7.0. Experiments were conducted using this initial pH only. For both parameters, very high removal efficiencies were achieved rapidly (6 min) and with low current values. Further RO studies conducted using EC-pretreated water proved that all fouling indicators such as flow, pressure drop, and silt density index (SDI) showed less fouling when EC was added prior to EO [108].

Mineral treatment processes produce wastewater containing suspended and stable colloidal particles which degrade recirculation of water in processing plants. Such synthetic aqueous solutions containing quartz were treated by batch-EC using Al/SS electrodes. The median particle size of the quartz-in-water (320 mg/l quartz, initial pH 4) was 11.61 μm. A comparison between EC and CC (using aluminum sulfate in jar tests) was made, achieving similar removal efficiencies (around 90%) when similar amounts of aluminum were added to the water. The optimum pH range of CC was found to be 6 - 9, which was wider than that of EC. However, CC was found to acidify the water, whereas EC treatment shifted the initial suspension pH towards neutral. A 10-min treatment was sufficient for both methods and the kinetics of EC could be modeled with a second-order rate equation. No clearly superior treatment method for the wastewater in question could be proposed within the scope of the study, as no economic values were presented [109].

In another study, EC-treatment of synthetic wastewater was carried out in a batch electrochemical cell equipped with iron electrodes. Bentonite suspensions (~20 NTU turbidity) represented colloid-polluted wastes, as clays behave like hydrophobic colloids in water. Turbidity removals of 80% - 85% could be obtained with very low currents (4 A/m^2, 40-min EC-run). The effect of initial pH was explored with values of 3, 7, and 12, and only the neutral initial pH gave poor results. This was explained as being due to different destabilization mechanisms being prevalent in medias of different pH values. In acidic media, charge neutralization was considered to be the main removal mechanism, whereas sweep flocculation would be dominant in an alkaline solution of this type. It was concluded that the EC process can be applied to treatment of wastes polluted with colloids [110].

Surface Water and other Natural Water

Table 6 presents a summary of recent applications of EC in the treatment of surface water and other natural water.

Treating simulated surface water containing algae (one of the most dominant cyanobacteria, Microcystis aeruginosa) by batch -EC was studied. The initial cell density used in the experiments was maintained at 1.2×10^9 - 1.4×10^9 cells/l. Aluminum was found to be an excellent electrode material for this application compared with iron (no coloration of water and substantially greater removal efficiency). Interestingly, it was found that algae removal was accelerated dramatically with increased water temperature. Ultimately, complete algae removal was achieved with low values of current density and EEC: 10 A/m^2 and 0.4 kWh/m^3, respectively. Thus, the results were proposed to indicate the effectiveness of EC in algae removal, from both the technical and economic points of view [111].

Laboratory experiments were carried out to investigate the effectiveness of disinfection by EC in a batch mode using artificial wastewater containing Escherichia coli. Real north-Algerian dam water from Ghrib (known for having high hardness) and Keddara (high algae content) dams were also used. The EC parameters presented for the three waters in Table 6 are in the same order as mentioned above. Aluminum electrodes were found slightly more efficient than ordinary steel (Fe) and stainless steel electrodes. Electrochemical disinfection was proven effective, because the treatment times were rather low and total disinfection and algae removal were achieved [112].

In another study (not presented in Table 6), Al -EC in a batch mode was found to be a suitable process for decreasing hardness and removing bacteria, algae, and bacterial nutriments from two different raw surface waters. The water samples originated from a river and a pond. Complete disinfection was achieved [113].

Reduction of humic acids (HA) from 1000 mg/l synthetic solutions by a batch Al-EC system was studied [114]. The effect of applying electromagnetic (EM) treatment prior to EC was also investigated in both batch and continuous modes, of which the latter was found more suitable. EM is an attractively simple approach in which the water being treated flows through a magnetic field, and it consequently slightly changes some of its physicochemical properties. Both EM and EC processes were found to perform best at neutral pH. The 10-min EM-pretreatment was found to slightly further increase the removal efficiency of HA by EC from 96% to 100% [114].

Removal of NOM (natural organic matter) from surface water (inlet flow of a Finnish paper mill) by a batch EC process using aluminum electrodes has been studied using RSM and ANOVA. In modeled optimum conditions, the applied current density, treatment time, and EEC-value were low, with simultaneous high removal efficiency. A slight lowering of initial pH (naturally 5.8) was found beneficial.

Table 6: Recent applications of EC in the treatment of surface water and other natural water

Water and wastewater types used	Genuine (G)/ Synthetic (S) water	Anode/ cathode material	Reactor type	Volume treated [ml]	Optimum electrode gap [mm]	Optimum current density, treatment time and initial pH [A/m²], [min], []			Initial pollutant levels [mg/l]	Optimum removal efficiency [%]	Optimum EEC [kWh/m³]	Optimum EEC [kWh/kg]	Optimum OC [€/m³]	Optimum OC [€/kg]	Research group
Water containing algae (cyanobacteria)	S	Al[b] Fe	Batch	1000	10	10	45	4 - 7	Cyanobacteria: 1.2 × 10⁹ - 1.4 × 10⁹ [cells/l]	Cyanobacteria: 100	0.4	n.d.	n.d.	n.d.	Gao et al 2010 [111]
Dam waters	S + G + 2	Al[b] Steel SS	Batch	500	50	(15/0.8/0. 25) [A] = -(202/16 2/51)[g]	10[g] 35/ 30	7- 9.5	E. coli: n.d./n.d./ n.d.	E. coli: 98[g]/100/100	~23.6[e]/16.8 /~4.7[g]	n.d.	n.d.	n.d.	Ghernaout et al 2008 [112]
Water containing humic acids (HA)	S	Al	Batch Continuous (EM) + Batch (EC)	500	40	33.3	30 10[g] (0.01 l/min) + 30	7 7 + 7	HA: 1000	HA: 96 100	n.d	n.d	n.d.	n.d.	Ghernaout et al 2009 [114]
Surface water (river) containing a high concentration of NOM (paper mill inlet flow)	G	Al	Batch	500[f]	10[f]	4.8	12[f]	4.3	DOC: 18.35 UV 254 nm: 0.64 [absorbance] Turbidity: 0.51 [NTU]	DOC: 80 UV₂₅₄: 91 Turbidity: n.d.	0.4	n.d.	n.d	n.d.	Vepsäläinen et al 2009 [115]
Micro-polluted raw water	G	Al[b] Fe	Batch	1000	10	50	20	5 7.5 [7.5]	TOC: 8.5 - 16.2 Oil: 0.8 - 1.5 NH₃-N: 0.75 - 1.26	TOC: 70 Oil: 86 NH₃-N: 75	n.d	n.d	n.d	n.d	Li et al 2008 [116]
Marine water containing microalgae (for biodiesel production)	S + 2	SS	Batch	300	48	10 [V]	15	4 - 9	Microalgal: 600/300	Microalgal: 98/99	n.d	4.44-9.16 kWh/ kg dry microalgae	n.d	n.d	Uduman et al 2011 [117]
Freshwater and marine water containing microalgae (for biodiesel production)	S + 2	Al/h₂O₂ TiO₂[b] Fe/IrO₂ TiO₂)	Batch	1000	44	(6-15)[g]	(30/ 30)[g]	4 - 6 [4]	Microalgal: 300 - 600; 300 - 600	Microalgal: (80-95)[g]	n.d	~0.3 ~(1.5 - 2.0) kWh/ kg dry microalgae	n.d	n.d	Vandamme et al 2011 [118]
Pesticide contaminated (metribuzin, MB) ground water	S	Fe/SS	Batch Batch + UV	1300[f]	n.d	18	80	5 - 6 [6²] 6 - 7 [6²]	MB: 50 - 300 (optimum 200)	MB: 85 95	n.d	n.d	n.d	n.d	Yahiaoui et al 2011 [119]
Geothermal waters containing boron	G	Al	Batch	1500	5	15[b] 30[b] 60	30	8	B: 24	B: 73[b]-84[b]-96 12.8	0.73[b] - 2.5[b]	n.d	n.d	n.d	Yilmaz et al 2008 [120]
Riverwater containing mercury(II)	S + 2	Al Fe[b]	Batch	100	30	125	15	5 - 7 [7]	Hg²⁺: 100 Hg²⁺: 4 COD: 378	Hg²⁺: 100 Hg²⁺: 100 COD: 90	n.d	n.d	n.d	n.d	Nanseu-Njiki et al 2009 [121]

ᵃ= HRT (hydraulic retention time) in EC systems with continuous mode of operation [min]; ᵇ= Observed as the best electrode configuration of those tested; ᶜ= The natural, unmodified pH value of the water or wastewater (found optimal); ᵉ= Approximation calculation based on values given in the article at issue; ᶠ= Reactor volume (sample volume not mentioned); ᵍ = Optimum value estimated from the data in the article (precise value not given); ʰ= Additional "optimum value" estimated from the data in the article; ⁱ= Queried from the author; n.d. = Not determined.

The above values are presented in Table 6 for water at room temperature (23°C). The water samples taken from the river were at 3°C and EC -runs were also performed with water of this temperature. The removal efficiency was then only four percentage points lower, thus it was concluded that EC is a feasible treatment process for removal of NOM also during the cold water period in the Nordic countries [115].

Batch-EC was used and found feasible for treating micro-polluted surface water in laboratory-scale experiments. Aluminum was selected as the electrode material, because although iron produced nearly similar removal efficiencies, it also colored the water (to greenish at first and then to brown). Initial pollutant concentrations were very low (see Table 6), but so were also the optimal current density and treatment times (50 A/m^2 and 20 min, respectively). Also, no pH adjustment was found necessary [116].

Chlorococcum sp. and Tetraselmis sp. (results in this order in Table 6) were cultivated to produce artificial marine water to be separated by batch-EC from microalgae for use in biodiesel production. Both species were quite different but had approximate cell sizes of 10 μm. Altering the initial pH value between 4 - 9 had no clear impact on removal efficiencies; EC was found applicable over the whole scale tested. Both waters were naturally within this pH range: the first water had a pH value of 9.1 and the latter, 8.3. High recovery efficiencies were obtained, up to 98% and 99% for Chlorococcum sp. and Tetraselmis sp., respectively. Microalgae flotation due to hydrogen bubble attachment was documented using a hi-speed camera; this is presented in Figure 6 [117].

In another similar study, using EC to harvest synthetic marine water (containing Phaeodactylum tricornutum) and freshwater (containing Chlorella vulgaris) for biodiesel production was evaluated (results in this order in Table 6). Using Al as anodes was found clearly more efficient than using Fe. Both electrode configurations had cathodes made of IrO$_2$/TiO$_2$, which is fairly uncommon. The aluminum content of the harvested microalgal biomass was less than 1%, while the aluminum concentration of the process water was below 2 mg/L for C. Vulgaris and below 0.5 mg/l for P. tricornutum. Rather rarely, the effect of stirring the water was tested, within a range of 0 - 200 rpm. It was found that increasing the stirring speed significantly increased the performance of the EC process up to a value of 150 rpm, enhancing contact rates between coagulants and microalgal cells. However, further increasing the stirring rate was found to decrease the performance of the EC process to near the level of 0 rpm. This was proposed to be due to the break-up of flocs because of the high shear forces applied. Under optimal conditions, the EEC-values were around 0.3 kWh/kg$_{algae}$ harvested for P. tricornutum and approximately 1.5 - 2.0 kWh/kg$_{algae}$ for C. vulgaris, while the respective microalgal recovery rates were 80% and 95%. In specific triplication tests, the process was found to be repeatable. Compared to centrifugation, EC was thus suggested as substantially more energy-efficient. Finally, it was concluded that EC is a promising technology for harvesting marine microalgae, but tests with large-scale pilot EC reactors need to be done to confirm this [118].

A batch EC system using a Fe/SS electrode configuration was used to treat model pesticide-contaminated (metribuzin, MB) groundwater. Metribuzin is considered a general-use pesticide which belongs to the group of triazinone herbicides, and it is highly water-soluble (1.05 g/l). The initial MB values used in the study (50 - 300 mg/l) were similar to those measured in discharges from MB manufacturers. The performance of the EC process was compared with that of a batch combined EC + UV (ultraviolet) process, and the latter was found slightly more effective (MB removal efficiencies of 89% and 95%, respectively). Using a batch EF + UV process to treat the water was also investigated, but it resulted in lower removal efficiencies than the EC + UV process, and thus the results are not shown in Table 6. When the water was treated with UV alone for 80 min (optimal value found for EC + UV), approximately 12% removal efficiencies were achieved (seeFigure 6 in the original paper). The natural initial pH value (6) of the water was found optimal for both the EC and EC + UV processes, while low pH values would have been needed for optimal functioning of the EF + UV process. In the end, it was proposed that the process studied may be employed successfully to remove pesticides from water [119].

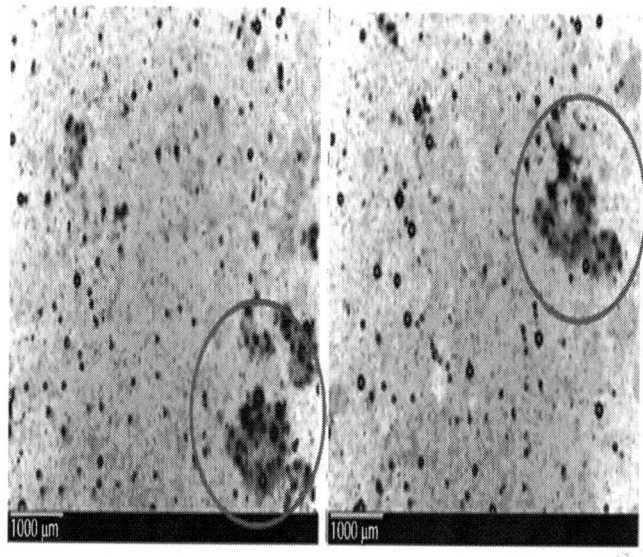

(a)

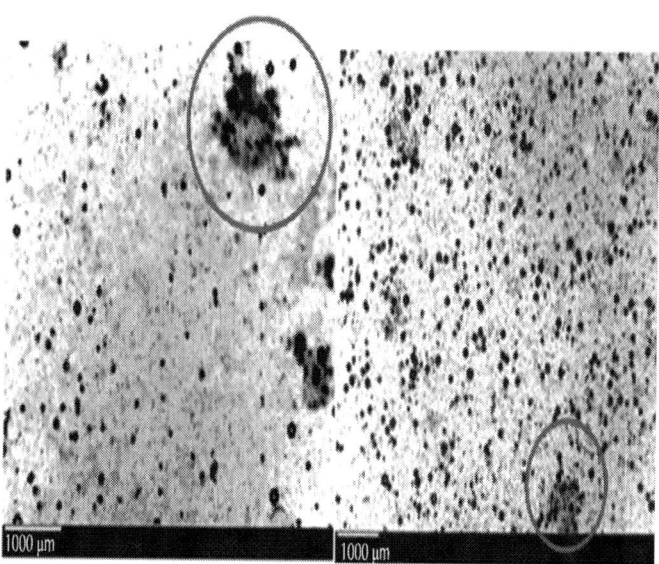

(b)

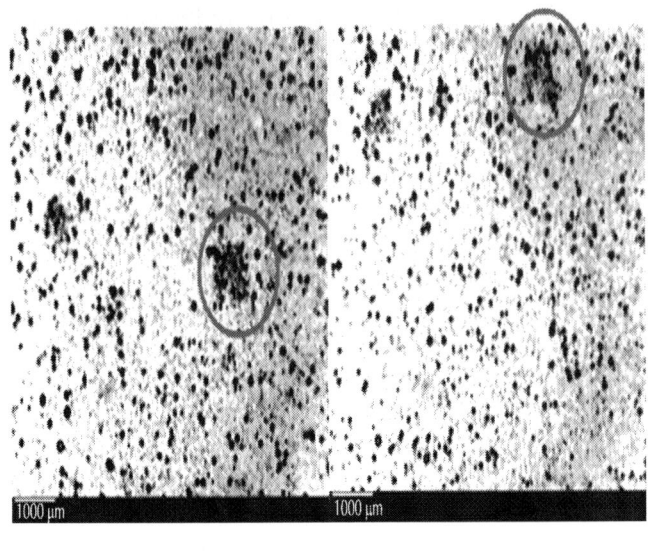

(c)

Figure 6: Hydrogen bubble-floc interaction at different times: For Chlorococcum sp.: (a) 0 s; (b) 0.4 s; and (c) 0.8 s after the current was turned off. For Tetraselmis sp.; (d) 0.4 s; (e) 0.7 s; and (f) 0.8 s after the current was turned off. (Main bubble-microalgae floc aggregates are circled in each figure). Adapted from [117].

Although boron is a vital micronutrient for plant and crop growth, high boron concentrations in irrigation water are known to be detrimental to them. Therefore, treatment of genuine geothermal water containing 24 mg/l boron was studied with an EC system working in a batch mode and using aluminum electrodes. An initial pH value of 8 was found optimal, which was near to the natural value of 6.5. It was found that decreasing the current density value from 60 A/m² (proposed as optimal) to 30 A/m² or 15 A/m² corresponds to a tremendous decrease in the EECvalue from 12.8 kWh/m³ to 2.3 kWh/m³ and 0.73 kWh/m³, respectively, while boron removal efficiencies decreased from 96% to 84% and 73 %, respectively. It was concluded that after the EC process, the effluent water could be used for irrigation [120].

Studies on mercury (II) removal from water were conducted using a laboratory-scale (100 ml) batch-EC system. A synthetic mercury-containing (4 mg/l) solution was first used to optimize the process. Iron was found to perform better than aluminum; however, both electrodes achieved very high removal efficiencies over a wide pH scale. Mercury was then added to a river water sample achieved very high removal efficiencies over a wide pH scale. Mercury was then added to a river water sample and the solution was treated in the previously determined optimum process conditions (see Table 6). Complete mercury removal was achieved also from the semi-synthetic mercury-contaminated river water, with 90% COD removal, as well. It was concluded that EC can be effective in the treatment of water polluted by mercury(II) ions [121].

Model Water and Wastewater Containing Heavy Metals, Nutrients, Cyanide and other Elements and Ions

EC studies presented in this category of waters can be simplified to have been conducted using small-scale laboratory Al, Fe, or SS batch reactors and synthetic model water or wastewater with only one particular pollutant removed (see Table 7). Therefore, not all of the studies [13,122-137] are discussed in the following text (although the optimum conditions found are presented in Table 7), but only those which differ notably from this in some way. Table 7 presents a summary of recent applications of EC in the treatment of waters containing heavy metals, cyanide, and other elements and ions.

Arsenic removal batch-EC experiments using iron were conducted in a laboratory with synthetic solutions, providing complete As(V) removal with a short treatment time. The experiments were further expanded to field tests in which 50-l batches of real arsenicand phosphate-containing groundwater were treated. In the field tests, a filtration step was also added after the EC, raising the removal efficiency of total arsenic from 97% to 100%. The economic values were found to be very low and the naturally neutral initial pH of the water was found optimal [122].

Iron-containing (25 mg/l) synthetic solutions were treated by batch-EC. Magnesium was used as the anode material (iron as the cathode), which is rare, but the EC system performed well and in optimum conditions very high removal efficiencies with very low current density values were achieved (see Table 7) [123].

Magnesium was used as the anode material (with a SS cathode) in yet another study. The synthetic water being treated contained boron and the batch EC system performed well in removing it. A scale-up batch EC system with an 8.5-l cell volume was also built and tested. It was found to produce exactly similar removal efficiencies with similar (optimum) process conditions (86% of initial boron removed at an initial pH of 7, using a low current density of 20 A/m^2). This was concluded to show the robustness of the EC process [125].

Removal of cadmium from aqueous solutions was studied with Al-EC in a batch mode. In addition to using a regular DC power source, application of an alternating pulse current (APC) to prevent passivation of the electrodes was studied. EC in the APC mode was found to perform slightly better than in the DC mode, with nearcomplete cadmium removal and a significantly shorter treatment time required. Also, a pilot-scale batch-EC system with a 2000-l cell volume was built and tested in this study. The results were consistent with the results obtained from the laboratory scale, showing that the process was technologically feasible and scalable. Adding high concentrations of co-existing ions (carbonate, phosphate, silicate, or arsenate) to the solution was found to decrease cadmium removal efficiency significantly due to competition for adsorption binding sites [126].

Removal of chromium from synthetic solutions with concentrations of 50, 100, 500, and 1000 mg/l (values given in this order in Table 7) were treated with a batchEC system using aluminium/iron electrodes. Use of APC was also studied in addition to the DC mode. During these experiments the polarity of the anode and cathode was thus changed every 4 min. It was found that the APC mode was more efficient than the DC mode here, also, with operating times 3%, 6%, 15%, and 25% shorter when treating initial Cr(VI) concentrations of 50, 100, 500, and 1000 mg/l, respectively. This makes the APC mode more cost-effective. Also, the turbidity values of the treated water were 1 NTU and 20 NTU with APC and DC, respectively. NaCl, KCl, PAC, and NaNO$_3$ were tested as a supporting electrolyte; NaCl and KCl were found most suitable in every aspect. In optimum conditions, complete removal of chromium was achieved [127].

Solutions containing radioactive strontium were treated with a batch-EC system. Initial concentrations of up to 100 mg/l were investigated.

Table 7: Recent applications of EC in the treatment of waters containing heavy metals, cyanide, and other elements and ions

Water and wastewater types used	Genuine (G)/ Synthetic (S) water	Anode/ cathode material	Reactor type	Volume treated [ml]	Optimum electrode gap [mm]	Optimum current density, treatment time and initial pH [A/m²], [min], []			Initial pollutant levels [mg/l]	Optimum removal efficiency [%]	Optimum EEC [kWh/m³]	Optimum EEC [kWh/kg]	Optimum OC [€/m³]	Optimum OC [€/kgx]	Research group
Contaminated groundwater (As)	S	Al/ Fe^b	Batch	650	13	2.5	2.5	5.5 - 7.5 [6.5]	As: 0.15	As: 99	0.014	n.d.	0.0047	n.d.	Kobya et al. 2011 [13]
Contaminated drinking water (As)	S G	Fe	Batch Batch (scale-up/ field test)	1000 50000	20 5	0.022 [A] = 3.9^a 2 [A]	(5/15)^g 180	5 - 7 [7^c]	As(V): 0.1 - 1.0 As$_{total}$: 0.45 - 0.67 Phosphate: 0.18 - 0.75	As(V): (~97/100)^g As$_{total}$: 100 Phosphate: 100	0.5 0.72 - 0.78	n.d.	n.d. 0.077	a.d.	Wan et al. 2011 [122]
Contaminated drinking water (Fe)	S	Mg/Fe	Batch	900	5	2^h/6	35	6	Fe: 5 - 25 (optimum 25)	Fe: 92^h/98	n.d.	n.d.	n.d.	n.d.	Vasudevan et al. 2009 [123]
Contaminated drinking water (Fe)	S	Al	Batch	1000	5	12.5	40	8	Fe: 2 - 15 (optimum 10)	Fe: 99	~9.4^g	n.d.	0.21	n.d.	Ghosh et al. 2010 [124]
Drinking water containing boron	S	Mg/SS	Batch	900	5	20	30	7	B: 3 - 7 (optimum 5)	B: 86	n.d.	n.d.	n.d.	n.d.	Vasudevan et al. 2010 [125]
Water containing cadmium	S	Al-alloy	Batch	1000^f	5	10^h/20	30 (AC) 45 (DC)	6 - 8 [7]	Cd: 10 - 50 (optimum 20)	Cd: 94^h/98 (AC) 92^h/96 (DC)	0.227^h/0.4 54 (AC) 0.881^h/1.0 0.2 (DC)	n.d.	n.d.	a.d.	Vasudevan et al. 2011 [126]
Water containing Cr(VI)	S	Fe/Al - Al/Fe (APC)	Batch	700^f	15	56/153/ 153/222	20/25/ 55/110	3 - 5 [5^c]	Cr(VI): 50/100/ 500/1000	Cr(VI): 98/98/ 99/100	4.0/16.3/ 20.2/58.0	n.d.	n.d.	a.d.	Keshmirizadeh et al. 2011 [127]
Water containing copper	S	Al	Batch	560	15	(9.1/ 9.0)^g [V]	(10.4/ 10.2)^g	6 - 8	Cu: 2.5 - 32.5 (optimum 14.2/15.0)^g	Cu: (81/90)^g	(4.07/ 6.32)^{e,g}	n.d.	n.d.	a.d.	Bhatti et al. 2011 [128]
Water containing fluoride	S	Al	Batch	2000	30	111	25	7	Fluoride: 25 - 125 (optimum 25)	Fluoride: 95	n.d.	n.d.	0.90 - 1.05	n.d.	Behbahani et al. 2011 [129]
Aqueous solution containing indium	S	Fe/Al	Batch	500	20	64	50/60^h 108^h	2.3^d	In: 20 - 80 (optimum 20-40^h/80^h)	In: 90/90^h/90^h	~0.002^h/ n.d./n.d.	0.085/n.d. /n.d.	n.d.	n.d.	Chou & Huang 2009 [130]
Wastewater containing manganese	S	Al	Batch + Aerator	250	10	62.5	(30/ 60)^g	7	Mn^{2+}: 25 - 400 (optimum 100)	Mn^{2+}: (78/94)^g	n.d.	n.d.	n.d.	n.d.	Shafaei et al. 2010 [131]
Water containing phosphate	S	Al^b Fe	Batch	5000	9	(1/10/ 30/50)^g	~(380/ 150/ 80/25)^g	7	Phosphate: 27	Phosphate: (100/100/ 100/100)^g	0.06/~0.3 8^g ~0.58^g/ 0.73	n.d.	n.d.	n.d.	Lacasa et al. 2011 [132]
Phosphate-contaminated water	S	Al/SS Al-alloy /SS^b Fe/SS	Batch	900	5	20	30	6 - 9 [7]	Phosphate: 100	Phosphate: 99	n.d.	n.d.	n.d.	n.d.	Vasudevan et al. 2009 [133]
Water containing strontium	S	Al SS^b	Batch	200	60	80	50	5 - 7 [7]	Sr(II): 10 - 100 (optimum 10)	Sr(II): 93	n.d.	n.d.	~2^g	n.d.	Murthy & Parmar 2011 [134]
Cyanide-laden wastewater	S	Al Al-Fe Fe Fe-Al^b	Batch Continuous	250 250	30 30	150	(20/30)^g 140^a (0.00671 /min)	9.5^d	Cyanide: 300	Cyanide: (98/100)^g 100	n.d.	n.d.	n.d.	n.d.	Moussavi et al. 2011 [135]
Contaminated soil (Pb, Zn, Cd) washing solution containing EDTA	S	Al/SS	Batch	500	10	160^g	30^g	7.1^d	Pb: 1220 - 1620 Zn: 230 - 290 Cd: 8 - 10 EDTA 16500 - 19700	Pb: 95 Zn: 68 Cd: 66 EDTA: ~50	~8^g	n.d.	n.d.	n.d.	Lestan & Pociecha 2010 [136]
Aqueous solution containing tannic acid (TA)	S	Fe	Batch	200	20	100	75^d	7 - 9 [9]	TA: 50	TA: 99	n.d.	2.4 kWh/h/kg$_{COD}$	n.d.	n.d.	Mansouri et al. 2011 [137]

[a]= HRT (hydraulic retention time) in EC systems with continuous mode of operation [min]; [b]= Observed as the best electrode configuration of those tested; [c]= The natural, unmodified pH value of the water or wastewater (found optimal); [d]= The natural, unmodified pH value of the water or wastewater (the effect of pH not researched); [e]= Approximation calculation based on values given in the paper; [f] = Reactor volume (sample volume not mentioned); [g]= Optimum value estimated from the data in the article (precise value not given); [h]= Additional "optimum value" estimated from the data in the article; n.d. = Not determined.

Even though according to Figure 5 in the article and as mentioned in the text, a neutral initial pH value was found to be the most efficient, nonetheless pH 5 was chosen as the optimum by the authors. However, the difference between these values was negligible. Raising the temperature above 30°C was found to decrease efficiency. Stainless steel was found to be a clearly better electrode material than Al for this application, with optimal removal efficiency of 93%. In this study, removal efficiency was found to greatly improve when the distance between the electrodes was increased [134].

Synthetic wastewater containing 300 mg/l cyanide was treated by EC operating in both batch and continuous modes. Out of the four electrode combinations tested, Fe/Al was found the best-performing, with Fe/Fe nearly as effective. When Al was used as the anode material, substantially lower removal efficiencies for cyanide were achieved. Both the batch and continuous EC experiments were conducted using additional aeration in the EC reactor. Aeration was found to improve the efficiency of cyanide removal in the 30-min batch EC test by approximately 6 percentage points, raising it from 94% to 100%. Cyanide could be removed completely from the wastewater with EC using both operating modes, although the batch mode was found more efficient (see Table 7). Therefore, EC was concluded to be a promising technique for treating cyanide-laden wastewater [135].

DISCUSSION

Research on various applications of EC has been conducted extensively around the world recently. A large number of these studies have been conducted in the Middle Eastern countries and India. In these studies EC has been found to be a feasible, economical and ecological alternative in the treatment of various types of water and wastewater with promising results. The interest in EC seems to be on the rise. Apart from a handful of articles, this paper discusses EC literature published in 2008- 2011.

A few of the EC systems studied worked in a hybrid treatment mode (EM, EF, addition of polymer, UV-light, aeration, filtration). RSM has been successfully applied by several authors to optimize the EC process, providing high levels of significance and very low percentages of experimental error (related to theoretical models) in the papers discussed in this study. Therefore, RSM could be applied with EC to find case-specific optimum operating conditions. Duplication or triplication of all test runs was performed in a number of papers. They confirmed that the EC process is repeatable with low (a few percent) experimental error.

Innovations such as collecting the hydrogen gas produced during EC and utilizing it for the EC process's own energy demand (leading to a 13% decrease in the energy costs of the EC process), solar-powered EC systems, partially recirculating EC sludge supernatant, applying sonar or magnetic field treatment prior to EC (inducing changes in the structure of the studied aqueous solutions, which in turn leads to enhancement of the EC process by a few percentage points), and applying APC have been studied, with promising results.

Most of the authors have conducted their EC studies using small (250 - 2000 ml) laboratory-scale reactors with magnetic stirring and virtually all of the EC systems operated in a batch mode. Systems working in a continuous mode and larger scale-up systems have also been investigated, showing promising treatment results and the scalability of the process. Such EC systems should be applied more in future studies.

Slightly over half of the EC studies scrutinized in this paper were carried out using genuine water or wastewater. In a few studies both genuine and synthetic wastewaters modeling a certain similar type of wastewater were used. The wastewater from paper the industry used by the authors was all genuine. Nearly all of the wastewaters from the oil and food industries were also genuine. Synthetic wastewaters have been used much more extensively in the EC studies of tannery, textile, and colored wastewaters, as only a third of such studies discussed here were done using real wastewater. The other types of industrial wastewater studied were mostly genuine.

In nearly all of the studies discussed here, the electrode materials used were made of aluminum, iron, or SS in different combinations, with only a few exceptions. As it can be seen from Tables 1-7, the superiority of different electrode materials seems to vary between different types of aqueous solutions being treated and must therefore always be researched case-specifically. In some of the studies, aluminum and iron performed so similarly that a clear choice of the superior material could not be suggested. It must also be taken into consideration that treatment costs and efficiencies are not always the sole factors when choosing between different electrode materials to be used, because other technical aspects (e.g. floc properties, coloration of water by iron, etc.) may also affect the decision-making.

The distance between the electrodes fluctuated between 2 - 70 mm; however, most setups used by the authors considered here had as electrode gap of 5 - 20 mm. The effect of this parameter on the feasibility of the EC treatment was rarely studied; e.g. [10,29,124,134]. This also applies to the effect of temperature and stirring, although all of these have been shown to have a varied effect on the removal efficiencies of EC [31,48,115,117, 120,134,138].

Even though the EC process seems to function well over a wide range of pH values in most studies, generally a relatively narrow pH range (depending on the electrode materials used) where the process performed optimally could be found. This pH range was mostly found to be close to neutral pH values, as observed in Tables 1-7.

The treatment costs and electricity consumption of the EC process in optimal process conditions were not calculated and presented by all the authors, but the aforementioned values were mostly somewhat low when they were presented (typically around 0.1 - 1.0 EUR/m^3 and 0.4 - 4.0 kWh/m^3, respectively, see Tables 1-7). On the whole, the values for optimal treatment costs and electricity consumption varied greatly between different studies and different types of aqueous solutions, the aforementioned values fluctuating between 0.0047 - 6.74 EUR/m^3 and 0.002 - 58.0 kWh/m^3, respectively. However, both values were close to the lower end of the scale presented above.

In most of the papers reviewed in chapter 3.1 the EEC -values were given, most of them being inside or close to a range of 0.1 - 0.8 kWh/

m^3. Operating costs were calculated in only two of the papers, however, they were found to be similar in degree: 0.12 and 0.24 EUR/m^3, which can be considered very low. Oily wastewaters had economic numbers close to each other. As for OC, 0.2 - 0.4 EUR/m^3 seemed to be an average value, and correspondingly, 2 - 6 kWh/m^3 was the average EEC -value. The results of a very recent investigation of EC treatment of bio oil-in-water and synthetic oil-in-water emulsions support these findings [139].

Food industry wastewater treated with EC had OC -values in a range of 0.051 - 4.1 EUR/m^3 (given in less than half of the papers). The papers presented in chapter 3.5 had economic numbers presented in about half of the papers and their fluctuation was considerable (between 0.062 - 6.74 EUR/m^3 and 0.14 - 19.8 kWh/m^3). This can be considered to be mainly due to the significant variance of water types, compositions, and their pollutant concentrations. Nearly all of the EC treatment results for waters presented in chapter 3.7 had very low OC-values when they were presented; the same is true for EEC-values (see Table 7). This could be mainly due to the fact the waters in this category were synthetic, modeling the removal of only one pollutant at a time and at mostly relatively low concentrations. The publications reviewed in chapters 3.2 and 3.6 had no OC-values given. Paper industry wastewater EEC-values were also absent, whereas in surface water category they fluctuated substantially (between 0.4 - 23.6 kWh/m^3).

The observed optimal current densities varied greatly, but in most studies they were found to be in the range of 10 - 150 A/m^2. When the various waters were divided into categories (chapters 3.1 - 3.7), it seems the optimal current density values of oily wastewaters were the highest of all the categories, on average (approx. 160 A/m^2). Waters from other industries also had optimal current density values often higher than average. The corresponding value for waters and wastewaters presented in chapters 3.6 and 3.7 were the lowest, on average: approx. 45 - 50 and 60 - 65 A/m^2, respectively. Both water categories had most of their optimal current values inside the range of 5 - 60 A/m^2. For wastewaters presented in chapter 3.5, optimal current density values were largely in the range of 100 - 150 A/m^2. These observations could be related to the high levels of pollutants in oily and other industrial wastewaters and, on the contrary, to surface waters and modeled waters and wastewaters being only mildly polluted in comparison, with relatively low concentrations of pollutants.

One of the advantages of the EC process is its fast treatment capability, and in a vast majority of the studies discussed in this paper, optimal treatment times were found to be in the range of 5 - 60 min (these figures do not take required sedimentation times into account, which are generally rather short). More than half of the authors found an optimum treatment time of 30 min or less. In some cases a two-to three-hour EC treatment time was needed, however some wastewater could be purified rapidly in a few minutes. As with other functional parameters, treatment duration seems to be strongly dependent on the type of water being purified and its concentration. The applied current density also has a significant impact on treatment duration.

CONCLUSIONS

The range of feasible EC applications is expanding. In a vast majority of the studies discussed in this paper, optimal treatment times were found to be in the range of 5 - 60 min (not taking into account required sedimentation times). More than half of the authors found an optimum treatment time of 30 min or less. Observed optimal current densities varied greatly, but in most studies they were found to be in the range of 10 - 150 A/m^2. Even though the EC process seems to function well over a wide range of pH values in most studies, generally a relatively narrow pH range could be found where the process performed optimally. This pH range was mostly found to be close to the neutral pH value.

The superiority of different electrode materials seems to vary between different types of aqueous solutions being treated, and must therefore always be studied casespecifically. Both OC and EEC-values were found to fluctuate greatly between different water types being treated, between 0.0047 - 6.74 EUR/m^3 and 0.002 - 58.0 kWh/m^3, but they were generally rather low (typically around 0.1 - 1.0 EUR/m^3 and 0.4 - 4.0 kWh/m^3, respectively). To conclude, EC has great potential in purification of various types of water and wastewater and seems to be a feasible and economical alternative in this field, although more research is needed, especially using larger-scale and/or continuous systems and focusing on the fundamentals of the EC process.

ACKNOWLEDGEMENTS

This study was supported by Maa—ja vesitekniikan tuki ry. and Academy of Finland (AquAlSi-project). Expression of gratitude is presented to Authorized Translator Keith Kosola for revising the language.

REFERENCES

1. P. K. Holt, G. W. Barton and C. A. Mitchell, "The Future for Electrocoagulation as a Localised Water Treatment Technology," Chemosphere, Vol. 59, No. 3, 2005, pp. 355-367. doi:10.1016/j.chemosphere.2004.10.023

2. G. Chen, "Electrochemical Technologies in Wastewater Treatment," Separation and Purification Technology, Vol. 38, No. 1, 2004, pp. 11-41.doi:10.1016/j.seppur.2003.10.006

3. M. M. Emamjomeh and M. Sivakumar, "Review of Pollutants Removed by Electrocoagulation and Electrocoagulation/Flotation Processes," Journal of Environmental Management, Vol. 90, No. 5, 2009, pp. 1663-1679. doi:10.1016/j.jenvman.2008.12.011

4. E. Butler, Y.-T. Hung, R. Y.-L. Yeh and M. S. Al Ahmad, "Electrocoagulation in Wastewater Treatment," Water, Vol. 3, No. 2, 2011, pp. 495-525.doi:10.3390/w3020495

5. M. Y. A. Mollah, R. Schennach, J. R. Parga and D. L. Cocke, "Electrocoagulation (EC)—Science and Applications," Journal of Hazardous Materials, Vol. 84, No. 1, 2001, pp. 29-41. doi:10.1016/S0304-3894(01)00176-5

6. P. R. Kumar, S. Chaudhari, K. C. Khilar and S. P. Mahajan, "Removal of Arsenic from Water by Electrocoagulation," Chemosphere, Vol. 55, No. 9, 2004, pp. 1245- 1252. doi:10.1016/j.chemosphere.2003.12.025

7. X. Chen, G. Chen and P. L. Yue, "Separation of Pollutants from Restaurant Wastewater by Electrocoagulation," Separation and Purification Technology, Vol. 19, No. 1-2, 2000, pp. 65-76. doi:10.1016/S1383-5866(99)00072-6

8. S. Zodi, O. Potier, F. Lapicque and J. Leclerc, "Treatment of the Textile Wastewaters by Electrocoagulation: Effect of Operating

Parameters on the Sludge Settling Characteristics," Separation and Purification Technology, Vol. 69, No. 1, 2009, pp. 29-36. doi:10.1016/j.seppur.2009.06.028

9.   S. Gao, M. Du, J. Tian, J. Yang, J. Yang, F. Ma and J. Nan, "Effects of Chloride Ions on Electro-CoagulationFlotation Process with Aluminum Electrodes for Algae Removal," Journal of Hazardous Materials, Vol. 182, No. 1-3, 2010, pp. 827-834.doi:10.1016/j.jhazmat.2010.06.114

10.   E. Terrazas, A. Vázquez, R. Briones, I. Lázaro and I. Rodríguez, "EC Treatment for Reuse of Tissue Paper Wastewater: Aspects That Affect Energy Consumption," Journal of Hazardous Materials, Vol. 181, No. 1-3, 2010, pp. 809-816.doi:10.1016/j.jhazmat.2010.05.086

11.   S. Kongjao, S. Damronglerd and M. Hunsom, "Simultaneous Removal of Organic and Inorganic Pollutants in Tannery Wastewater Using Electrocoagulation Technique," Korean Journal of Chemical Engineering, Vol. 25, No. 4, 2008, pp. 703-709. doi:10.1007/s11814-008-0115-1

12.   İ. A. Şengil and M. Özacar, "The Decolorization of C.I. Reactive Black 5 in Aqueous Solution by Electrocoagulation Using Sacrificial Iron Electrodes," Journal of Hazardous Materials, Vol. 161, No. 2-3, 2009, pp. 1369- 1376. doi:10.1016/j.jhazmat.2008.04.100

13.   M. Kobya, F. Ulu, U. Gebologlu, E. Demirbas and M. S. Oncel, "Treatment of Potable Water Containing Low Concentration of Arsenic with Electrocoagulation: Different Connection Modes and Fe-Al Electrodes," Separation and Purification Technology, Vol. 77, No. 3, 2011, pp. 283-293. doi:10.1016/j.seppur.2010.12.018

14.   G. Mouedhen, M. Feki, M. D. P. Wery and H. F. Ayedi, "Behavior of Aluminum Electrodes in Electrocoagulation Process," Journal of Hazardous Materials, Vol. 150, No. 1, 2008, pp. 124-135. doi:10.1016/j.jhazmat.2007.04.090

15.   K. Yetilmezsoy, F. Ilhan, Z. Sapci-Zengin, S. Sakar and M. T. Gonullu, "Decolorization and COD Reduction of UASB Pretreated Poultry Manure Wastewater by Electrocoagulation Process: A Post-Treatment Study," Journal of Hazardous Materials, Vol. 162, No. 1, 2009, pp. 120-132. doi:10.1016/j.jhazmat.2008.05.015

16. M. Rebhun and M. Lurie, "Control of Organic Matter by Coagulation and Floc Separation," Water Science and Technology, Vol. 27, No. 11, 1993, pp. 1-20.

17. M. Kobya, O. T. Can and M. Bayramoglu, "Treatment of Textile Wastewaters by Electrocoagulation Using Iron and Aluminum Electrodes," Journal of Hazardous Materials, Vol. 100, No. 1-3, 2003, pp. 163-178. doi:10.1016/S0304-3894(03)00102-X

18. L. D. Benefield, J. F. Judkins and B. L. Weand, "Process Chemistry for Water and Wastewater Treatment," Prentice-Hall, Englewood Cliffs, 1982.

19. A. J. Rubin, "Aqueous-Environmental Chemistry of Metals," Ann Arbor Science Publishers, Ann Arbor, 1974.

20. F. Akbal and S. Camci, "Comparison of Electrocoagulation and Chemical Coagulation for Heavy Metal Removal," Chemical Engineering & Technology, Vol. 33, No. 10, 2010, pp. 1655-1664. doi:10.1002/ceat.201000091

21. J. Duan and J. Gregory, "Coagulation by Hydrolysing Metal Salts," Advances in Colloid and Interface Science, Vol. 100-102, 2003, pp. 475-502. doi:10.1016/S0001-8686(02)00067-2

22. A. Kumar, "Tannery Effluent," In: M. Doble and A. Kumar, Eds., Biotreatment of Industrial Effluents, Butterworth-Heinemann, Burlington, 2005, pp. 133-143.

23. R. M. Makdisi, "Tannery Wastes Definition, Risk Assessment and Cleanup Options, Berkeley, California," Journal of Hazardous Materials, Vol. 29, No. 1, 1991, pp. 79- 96.doi:10.1016/0304-3894(91)87075-D

24. N. M. Mahmoodi, M. Arami, N. Y. Limaee and N. S. Tabrizi, "Decolorization and Aromatic Ring Degradation Kinetics of Direct Red 80 by UV Oxidation in the Presence of Hydrogen Peroxide Utilizing TiO_2 as a Photocatalyst," Chemical Engineering Journal, Vol. 112, No. 1-3, 2005, pp. 191-196. doi:10.1016/j.cej.2005.07.008

25. Y. Yalçın Şevki, "Optimization of Bomaplex Red CR-L Dye Removal from Aqueous Solution by Electrocoagulation Using Aluminum Electrodes," Journal of Hazardous Materials, Vol. 153, No. 1-2, 2008, pp. 194-200. doi:10.1016/j.jhazmat.2007.08.034

26. O. Apaydin, U. Kurt and M. T. Gönüllü, "An Investigation on the Treatment of Tannery Wastewater by Electrocoagulation," Global Nest Journal, Vol. 11, No. 4, 2009, pp. 546-555.

27. L. Szpyrkowicz, S. N. Kaul and R. N. Neti, "Tannery Wastewater Treatment by Electro-Oxidation Coupled with a Biological Process," Journal of Applied Electrochemistry, Vol. 35, No. 4, 2005, pp. 381-390. doi:10.1007/s10800-005-0796-7

28. S. Zodi, O. Potier, F. Lapicque and J. Leclerc, "Treatment of the Industrial Wastewaters by Electrocoagulation: Optimization of Coupled Electrochemical and Sedimentation Processes," Desalination, Vol. 261, No. 1-2, 2010, pp. 186-190. doi:10.1016/j.desal.2010.04.024

29. S. Aoudj, A. Khelifa, N. Drouiche, M. Hecini and H. Hamitouche, "Electrocoagulation Process Applied to Wastewater Containing Dyes from Textile Industry," Chemical Engineering and Processing: Process Intensification, Vol. 49, No. 11, 2010, pp. 1176-1182. doi:10.1016/j.cep.2010.08.019

30. C. Phalakornkule, P. Sukkasem and C. Mutchimsattha, "Hydrogen Recovery from the Electrocoagulation Treatment of Dye-Containing Wastewater," International Journal of Hydrogen Energy, Vol. 35, No. 20, 2010, pp. 10934-10943. doi:10.1016/j.ijhydene.2010.06.100

31. C. Phalakornkule, S. Polgumhang, W. Tongdaung, B. Karakat and T. Nuyut, "Electrocoagulation of Blue Reactive, Red Disperse and Mixed Dyes, and Application in Treating Textile Effluent," Journal of Environmental Management, Vol. 91, No. 4, 2010, pp. 918-926. doi:10.1016/j.jenvman.2009.11.008

32. P. Durango-Usuga, F. Guzmán-Duque, R. Mosteo, M. V. Vazquez, G. Peñuela and R. A. Torres-Palma, "Experimental Design Approach Applied to the Elimination of Crystal Violet in Water by Electrocoagulation with Fe or Al Electrodes," Journal of Hazardous Materials, Vol. 179, No. 1-3, 2010, pp. 120-126. doi:10.1016/j.jhazmat.2010.02.067

33. E. Z. El-Ashtoukhy and N. K. Amin, "Removal of Acid Green Dye 50 from Wastewater by Anodic Oxidation and Electrocoagulation—A Comparative Study," Journal of Hazardous Materials, Vol. 179, No. 1-3, 2010, pp. 113-119. doi:10.1016/j.jhazmat.2010.02.066

34. J. B. Parsa, H. R. Vahidian, A. R. Soleymani and M. Abbasi, "Removal of Acid Brown 14 in Aqueous Media by Electrocoagulation: Optimization Parameters and Minimizing of Energy Consumption," Desalination, Vol. 278, No. 1-3, 2011, pp. 295-302.doi:10.1016/j.desal.2011.05.040

35. A. Aleboyeh, N. Daneshvar and M. B. Kasiri, "Optimization of C.I. Acid Red 14 Azo Dye Removal by Electrocoagulation Batch Process with Response Surface Methodology," Chemical Engineering and Processing: Process Intensification, Vol. 47, No. 5, 2008, pp. 827-832. doi:10.1016/j.cep.2007.01.033

36. M. Y. A. Mollah, J. A. G. Gomes, K. K. Das and D. L. Cocke, "Electrochemical Treatment of Orange II Dye Solution—Use of Aluminum Sacrificial Electrodes and Floc Characterization," Journal of Hazardous Materials, Vol. 174, No. 1-3, 2010, pp. 851-858.doi:10.1016/j.jhazmat.2009.09.131

37. M. S. Secula, I. Creţescu and S. Petrescu, "An Experimental Study of Indigo Carmine Removal from Aqueous Solution by Electrocoagulation," Desalination, Vol. 277, No. 1-3, 2011, pp. 227-235. doi:10.1016/j.desal.2011.04.031

38. B. Merzouk, B. Gourich, A. Sekki, K. Madani, C. Vial and M. Barkaoui, "Studies on the Decolorization of Textile Dye Wastewater by Continuous Electrocoagulation Process," Chemical Engineering Journal, Vol. 149, No. 1-3, 2009, pp. 207-214.doi:10.1016/j.cej.2008.10.018

39. F. Zidane, P. Drogui, B. Lekhlif, J. Bensaid, J. F. Blais, S. Belcadi and K. El Kacemi, "Decolourization of DyeContaining Effluent Using Mineral Coagulants Produced by Electrocoagulation," Journal of Hazardous Materials, Vol. 155, No. 1-2, 2008, pp. 153-163. doi:10.1016/j.jhazmat.2007.11.041

40. A. R. Khataee, V. Vatanpour and A. R. Amani Ghadim, "Decolorization of C.I. Acid Blue 9 Solution by UV/ Nano-TiO_2, Fenton, Fenton-Like, Electro-Fenton and Electrocoagulation Processes: A Comparative Study," Journal of Hazardous Materials, Vol. 161, No. 2-3, 2009, pp. 1225-1233. doi:10.1016/j.jhazmat.2008.04.075

41. A. H. Essadki, M. Bennajah, B. Gourich, C. Vial, M. Azzi and H. Delmas, "Electrocoagulation/Electroflotation in an External-Loop Airlift Reactor—Application to the Decolorization of

Textile Dye Wastewater: A Case Study," Chemical Engineering and Processing: Process Intensification, Vol. 47, No. 8, 2008, pp. 1211-1223.doi:10.1016/j.cep.2007.03.013

42. B. Merzouk, M. Yakoubi, I. Zongo, J.-P. Leclerc, G. Paternotte, S. Pontvianne and F. Lapicque, "Effect of Modification of Textile Wastewater Composition on Electrocoagulation Efficiency," Desalination, Vol. 275, No. 1-3, 2011, pp. 181-186.doi:10.1016/j. desal.2011.02.055

43. A. Benhadji, M. Taleb Ahmed and R. Maachi, "Electrocoagulation and Effect of Cathode Materials on the Removal of Pollutants from Tannery Wastewater of Rouï- ba," Desalination, Vol. 277, No. 1-3, 2011, pp. 128-134. doi:10.1016/j.desal.2011.04.014

44. K. S. P. Kalyani, N. Balasubramanian and C. Srinivasakannan, "Decolorization and COD Reduction of Paper Industrial Effluent Using Electro-Coagulation," Chemical Engineering Journal, Vol. 151, No. 1-3, 2009, pp. 97-104. doi:10.1016/j.cej.2009.01.050

45. M. Zaied and N. Bellakhal, "Electrocoagulation Treatment of Black Liquor from Paper Industry," Journal of Hazardous Materials, Vol. 163, No. 2-3, 2009, pp. 995- 1000.doi:10.1016/j. jhazmat.2008.07.115

46. M. Uğurlu, A. Gürses, Ç. Doğar and M. Yalçın, "The Removal of Lignin and Phenol from Paper Mill Effluents by Electrocoagulation," Journal of Hazardous Materials, Vol. 87, No. 3, 2008, pp. 420-428. doi:10.1016/j.jenvman.2007.01.007

47. S. Zodi, J. Louvet, C. Michon, O. Potier, M. Pons, F. Lapicque and J. Leclerc, "Electrocoagulation as a Tertiary Treatment for Paper Mill Wastewater: Removal of NonBiodegradable Organic Pollution and Arsenic," Separation and Purification Technology, Vol. 81 , No. 1, 2011, pp. 62-68. doi:10.1016/j.seppur.2011.07.002

48. R. Katal and H. Pahlavanzadeh, "Influence of Different Combinations of Aluminum and Iron Electrode on Electrocoagulation Efficiency: Application to the Treatment of Paper Mill Wastewater," Desalination, Vol. 265, No. 1-3, 2011, pp. 199-205.doi:10.1016/j.desal.2010.07.052

49. M. Vepsäläinen, H. Kivisaari, M. Pulliainen, A. Oikari and M. Sillanpää, "Removal of Toxic Pollutants from Pulp Mill Effluents by Electrocoagulation," Separation and Purification Technology, Vol. 81, No. 2, 2011, pp. 141-150. doi:10.1016/j.seppur.2011.07.017

50. S. Bellebia, S. Kacha, A. Z. Bouyakoub and Z. Derriche, "Experimental Investigation of Chemical Oxygen Demand and Turbidity Removal from Cardboard Paper Mill Effluents Using Combined Electrocoagulation and Adsorption Processes," Environmental Progress & Sustainable Energy, Vol. 31, No. 3, 2011, pp. 361-370. doi:10.1002/ep.10556

51. M. Hanafy and H. I. Nabih, "Treatment of Oily Wastewater Using Dissolved Air Flotation Technique," Energy Sources, Part A: Recovery, Utilization and Environmental Effects, Vol. 29, No. 2, 2007, pp. 143-159. doi:10.1080/009083190948711

52. E. GilPavas, K. Molina-Tirado and M. Á. Gómez-García, "Treatment of Automotive Industry Oily Wastewater by Electrocoagulation: Statistical Optimization of the Operational Parameters," Water Science and Technology, Vol. 60, No. 10, 2009, pp. 2581-2588. doi:10.2166/wst.2009.519

53. M. Saeedi and A. Khalvati-Fahlyani, "Treatment of Oily Wastewater of a Gas Refinery by Electrocoagulation Using Aluminum Electrodes," Water Environment Research, Vol. 83, No. 3, 2011, pp. 256-264. doi:10.2175/106143010X12780288628499

54. P. Drogui, M. Asselin, S. K. Brar, H. Benmoussa and J.-F. Blais, "Electrochemical Removal of Organics and Oil from Sawmill and Ship Effluents," Canadian Journal of Civil Engineering, Vol. 36, No.3, 2009, pp. 529-539. doi:10.1139/L09-003

55. P. Cañizares, F. Martínez, C. Jiménez, C. Sáez and M. A. Rodrigo, "Coagulation and Electrocoagulation of Oil-inWater Emulsions," Journal of Hazardous Materials, Vol. 151, No. 1, 2008, pp. 44-51. doi:10.1016/j.jhazmat.2007.05.043

56. P. Cañizares, C. Jiménez, F. Martínez, M. A. Rodrigo and C. Sáez, "The pH as a Key Parameter in the Choice between Coagulation and Electrocoagulation for the Treatment of Wastewaters," Journal of Hazardous Materials, Vol. 163, No. 1, 2009, pp. 158-164.doi:10.1016/j.jhazmat.2008.06.073

57. C. J. Izquierdo, P. Cañizares, M. A. Rodrigo, J. P. Leclerc, G. Valentin and F. Lapicque, "Effect of the Nature of the Supporting Electrolyte on the Treatment of Soluble Oils by Electrocoagulation," Desalination, Vol. 255, No. 1-3, 2010, pp. 15-20.doi:10.1016/j.desal.2010.01.022

58. K. Bensadok, S. Benammar, F. Lapicque and G. Nezzal, "Electrocoagulation of Cutting Oil Emulsions Using Aluminium Plate Electrodes," Journal of Hazardous Materials, Vol. 152, No. 1, 2008, pp. 423-430. doi:10.1016/j.jhazmat.2007.06.121

59. M. H. El-Naas, S. Al-Zuhair, A. Al-Lobaney and S. Makhlouf, "Assessment of Electrocoagulation for the Treatment of Petroleum Refinery Wastewater," Journal of Hazardous Materials, Vol. 91, No. 1, 2009, pp. 180-185.doi:10.1016/j.jenvman.2009.08.003

60. [61] M. B. Agustin, W. P. Sengpracha and W. Phutdhawong, "Electrocoagulation of Palm Oil Mill Effluent," International Journal of Environmental Research and Public Health, Vol. 5 , No. 3, 2008, pp. 177-180. doi:10.3390/ijerph5030177

61. [62] T. J. Ahmad and D. Zawawi, "Palm Oil Mill Effluent (POME) Treatment by Using an Electrocoagulation Method," Research Journal of Chemistry and Environment, Vol. 15, No. 2, 2011, pp. 606-609.

62. [63] M. Tir and N. Moulai-Mostefa, "Optimization of Oil Removal from Oily Wastewater by Electrocoagulation Using Response Surface Method," Journal of Hazardous Materials, Vol. 158, No. 1, 2008, pp. 107-115. doi:10.1016/j.jhazmat.2008.01.051

63. [64] M. Asselin, P. Drogui, S. K. Brar, H. Benmoussa and J. Blais, "Organics Removal in Oily Bilgewater by Electrocoagulation Process," Journal of Hazardous Materials, Vol. 151, No. 2-3, 2008, pp. 446-455. doi:10.1016/j.jhazmat.2007.06.008

64. [65] U. Tezcan Un, A. S. Koparal and U. Bakir Ogutveren, "Electrocoagulation of Vegetable Oil Refinery Wastewater Using Aluminum Electrodes," Journal of Environmental Management, Vol. 90 , No. 1, 2009, pp. 428-433. doi:10.1016/j.jenvman.2007.11.007

65. [66] Y. Meas, J. A. Ramirez, M. A. Villalon and T. W. Chapman, "Industrial Wastewaters Treated by Electrocoagulation," Electrochimica Acta, Vol. 55, No. 27, 2010, pp. 8165-8171. doi:10.1016/j.electacta.2010.05.018

66. [67] F. Hanafi, O. Assobhei and M. Mountadar, "Detoxification and Discoloration of Moroccan Olive Mill Wastewater by Electrocoagulation," Journal of Hazardous Materials, Vol. 174, No. 1-3, 2010, pp. 807-812. doi:10.1016/j.jhazmat.2009.09.124

67. [68] M. Kobya, C. Ciftci, M. Bayramoglu and M. T. Sensoy, "Study on the Treatment of Waste Metal Cutting Fluids Using Electrocoagulation," Separation and Purification Technology, Vol. 60, No. 3, 2008, pp. 285-291. doi:10.1016/j.seppur.2007.09.003

68. [69] G. Moussavi, R. Khosravi and M. Farzadkia, "Removal of Petroleum Hydrocarbons from Contaminated Groundwater Using an Electrocoagulation Process: Batch and Continuous Experiments," Desalination, Vol. 278, No. 1-3, 2011, pp. 288-294.doi:10.1016/j.desal.2011.05.039

69. [70] K. Suehara, Y. Kawamoto, E. Fujii, J. Kohda, Y. Nakano and T. Yano, "Biological Treatment of Wastewater Discharged from Biodiesel Fuel Production Plant with Alkali-Catalyzed Transesterification," Journal of Bioscience and Bioengineering, Vol. 100, No. 4, 2005, pp. 437- 442. doi:10.1263/jbb.100.437

70. [71] O. Chavalparit and M. Ongwandee, "Optimizing Electrocoagulation Process for the Treatment of Biodiesel Wastewater Using Response Surface Methodology," Journal of Environmental Sciences, Vol. 21, No. 11, 2009, pp. 1491- 1496. doi:10.1016/S1001-0742(08)62445-6

71. [72] Y. Avsar, U. Kurt and T. Gonullu, "Comparison of Classical Chemical and Electrochemical Processes for Treating Rose Processing Wastewater," Journal of Hazardous Materials, Vol. 148, No. 1-2, 2007, pp. 340-345.doi:10.1016/j.jhazmat.2007.02.048

72. [73] B. Demirel, O. Yenigun and T. T. Onay, "Anaerobic Treatment of Dairy Wastewaters: A Review," Process Biochemistry, Vol. 40, No. 8, 2005, pp. 2583-2595.doi:10.1016/j.procbio.2004.12.015

73. [74] İ. A. Şengil and M. Özacar, "Treatment of Dairy Wastewaters by Electrocoagulation Using Mild Steel Electrodes," Journal of Hazardous Materials, Vol. 137, No. 2, 2006, pp. 1197-1205. doi:10.1016/j.jhazmat.2006.04.009

74. [75] B. Balannec, M. Vourch, M. Rabiller-Baudry and B. Chaufer, "Comparative Study of Different Nanofiltration and Reverse Osmosis Membranes for Dairy Effluent Treatment by Dead-End Filtration," Separation and Purification Technology, Vol. 42, No. 2, 2005, pp. 195-200. doi:10.1016/j.seppur.2004.07.013

75. [76] K. Bensadok, N. El Hanafi and F. Lapicque, "Electrochemical Treatment of Dairy Effluent Using Combined Al and Ti/Pt Electrodes System," Desalination, Vol. 280, No. 1-3, 2011, pp. 244-251. doi:10.1016/j.desal.2011.07.006

76. [77] S. Tchamango, C. P. Nanseu-Njiki, E. Ngameni, D. Hadjiev and A. Darchen, "Treatment of Dairy Effluents by Electrocoagulation Using Aluminium Electrodes," Science of the Total Environment, Vol. 408, No. 8, 2010, pp. 947-952. doi:10.1016/j.scitotenv.2009.10.026

77. [78] J. P. Kushwaha, V. C. Srivastava and I. D. Mall, "Organics Removal from Dairy Wastewater by Electrochemical Treatment and Residue Disposal," Separation and Purification Technology, Vol. 76, No. 2, 2010, pp. 198-205.doi:10.1016/j.seppur.2010.10.008

78. [79] M. Kobya, H. Hiz, E. Senturk, C. Aydiner and E. Demirbas, "Treatment of Potato Chips Manufacturing Wastewater by Electrocoagulation," Desalination, Vol. 190, No. 1-3, 2006, pp. 201-211. doi:10.1016/j.desal.2005.10.006

79. [80] D. Valero, J. M. Ortiz, V. García, E. Expósito, V. Montiel and A. Aldaz, "Electrocoagulation of Wastewater from Almond Industry," Chemosphere, Vol. 84, No. 9, 2011, pp. 1290-1295. doi:10.1016/j.chemosphere.2011.05.032

80. [81] G. Roa-Morales, E. Campos-Medina, J. Aguilera-Cotero, B. Bilyeu and C. Barrera-Díaz, "Aluminum Electrocoagulation with Peroxide Applied to Wastewater from Pasta and Cookie Processing," Separation and Purification Technology, Vol. 54, No. 1, 2007, pp. 124-129. doi:10.1016/j.seppur.2006.08.025

81. [82] M. Asselin, P. Drogui, H. Benmoussa and J. Blais, "Effectiveness of Electrocoagulation Process in Removing Organic Compounds from Slaughterhouse Wastewater Using Monopolar and Bipolar Electrolytic Cells," Chemosphere, Vol. 72, No. 11, 2008, pp. 1727-1733. doi:10.1016/j.chemosphere.2008.04.067

82. [83] L. J. Xu, B. W. Sheldon, D. K. Larick and R. E. Carawan, "Recovery and Utilization of Useful By-Products from Egg Processing Wastewater by Electrocoagulation," Poultry Science, Vol. 81, No. 6, 2002, pp. 785-792.

83. [84] M. Kobya and S. Delipinar, "Treatment of the Baker's Yeast Wastewater by Electrocoagulation," Journal of Hazardous Materials, Vol. 154, No. 1-3, 2008, pp. 1133-1140. doi:10.1016/j.jhazmat.2007.11.019

84. [85] J. K. Maghanga, F. K. Segor, L. Etiégni and J. Lusweti, "Electrocoagulation Method for Colour Removal in Tea Effluent: A Case Study of Chemomi Tea Factory in Rift Valley, Kenya,"

Bulletin of the Chemical Society of Ethiopia, Vol. 23, No. 3, 2009, pp. 371-381.doi:10.4314/bcse.v23i3.47661

85. [86] I. Linares-Hernández, C. Barrera-Díaz, G. Roa-Morales, B. Bilyeu and F. Ureña-Núñez, "Influence of the Anodic Material on Electrocoagulation Performance," Chemical Engineering Journal, Vol. 148, No. 1, 2009, pp. 97-105. doi:10.1016/j. cej.2008.08.007

86. [87] M. R. V. Lanza and R. Bertazzoli, "Cyanide Oxidation from Wastewater in a Flow Electrochemical Reactor," Industrial and Engineering Chemistry Research, Vol. 41, No. 1, 2002, pp. 22-26.

87. [88] M. Kobya, E. Demirbas, N. U. Parlak and S. Yigit, "Treatment of Cadmium and Nickel Electroplating Rinse Water by Electrocoagulation," Environmental Technology, Vol. 31, No. 13, 2010, pp. 1471-1481. doi:10.1080/09593331003713693

88. [89] I. Heidmann and W. Calmano, "Removal of Ni, Cu and Cr from a Galvanic Wastewater in an Electrocoagulation System with Feand Al-Electrodes," Separation and Purification Technology, Vol. 71, No. 3, 2010, pp. 308- 314. doi:10.1016/j. seppur.2009.12.016

89. [90] F. Akbal and S. Camcı, "Copper, Chromium and Nickel Removal from Metal Plating Wastewater by Electrocoagulation," Desalination,Vol. 269, No. 1-3, 2011, pp. 214-222.doi:10.1016/j. desal.2010.11.001

90. [91] I. Kabdaşlı, T. Arslan, T. Ölmez-Hancı, I. Arslan-Alaton and O. Tünay, "Complexing Agent and Heavy Metal Removals from Metal Plating Effluent by Electrocoagulation with Stainless Steel Electrodes," Journal of Hazardous Materials, Vol. 165, No. 1-3, 2009, pp. 838-845. doi:10.1016/j.jhazmat.2008.10.065

91. [92] M. Kobya, E. Demirbas, A. Dedeli and M. T. Sensoy, "Treatment of Rinse Water from Zinc Phosphate Coating by Batch and Continuous Electrocoagulation Processes," Journal of Hazardous Materials, Vol. 173, No. 1-3, 2010, pp. 326-334. doi:10.1016/j.jhazmat.2009.08.092

92. [93] C. Wang, W. Chou, L. Chen and S. Chang, "Silica Particles Settling Characteristics and Removal Performances of Oxide Chemical Mechanical Polishing Wastewater Treated by Electrocoagulation Technology," Journal of Hazardous

Materials, Vol. 161, No. 1, 2009, pp. 344-350. doi:10.1016/j.jhazmat.2008.03.099

93.   [94] C.-T. Wang and W.-L. Chou, "Performance of COD Removal from Oxide Chemical Mechanical Polishing Wastewater Using Iron Electrocoagulation," Journal of Environmental Science and Health—Part A Toxic/Hazardous Substances and Environmental Engineering, Vol. 44, No. 12, 2009, pp. 1289-1297. doi:10.1080/10934520903140090

94.   [95] M. Panizza and G. Cerisola, "Applicability of Electrochemical Methods to Carwash Wastewaters for Reuse. Part 2: Electrocoagulation and Anodic Oxidation Integrated Process," Journal of Electroanalytical Chemistry, Vol. 638, No. 2, 2010, pp. 236-240.doi:10.1016/j.jelechem.2009.11.003

95.   [96] M. Kumar, F. I. A. Ponselvan, J. R. Malviya, V. C. Srivastava and I. D. Mall, "Treatment of Bio-Digester Effluent by Electrocoagulation Using Iron Electrodes," Journal of Hazardous Materials, Vol. 165, No. 1-3, 2009, pp. 345-352.doi:10.1016/j.jhazmat.2008.10.041

96.   [97] D. Ryan, A. Gadd, J. Kavanagh, M. Zhou and G. Barton, "A Comparison of Coagulant Dosing Options for the Remediation of Molasses Process Water," Separation and Purification Technology, Vol. 58, No. 3, 2008, pp. 347- 352.doi:10.1016/j.seppur.2007.05.010

97.   [98] L. Zhang, G. Xue, N. Zhang, S. Liu, A. Duan and L. Wang, "Decolorization Study of Coking Wastewater by Continuous Electrocoagulation Process," ICMREE2011—Proedings 2011 International Conference on Materials for Renewable Energy and Environment, Shanghai, May 2011, pp. 860-863.

98.   [99] J. Zhu, F. Wu, X. Pan, J. Guo and D. Wen, "Removal of Antimony from Antimony Mine Flotation Wastewater by Electrocoagulation with Aluminum Electrodes," Journal of Environmental Sciences, Vol. 23, No. 7, 2011, pp. 1066-1071. doi:10.1016/S1001-0742(10)60550-5

99.   [100] F. Janpoor, A. Torabian and V. Khatibikamal, "Treatment of Laundry Waste-Water by Electrocoagulation," Journal of Chemical Technology and Biotechnology, Vol. 86, No. 8, 2011, pp. 1113-1120. doi:10.1002/jctb.2625

100. [101] C. Wang, W. Chou and Y. Kuo, "Removal of COD from Laundry Wastewater by Electrocoagulation/Electroflotation," Journal of Hazardous Materials, Vol. 164, No. 1, 2009, pp. 81-86. doi:10.1016/j.jhazmat.2008.07.122

101. [102] A. Akyol, "Treatment of Paint Manufacturing Wastewater by Electrocoagulation," Desalination, Vol. 285, 2012, pp. 91-99. doi:10.1016/j.desal.2011.09.039

102. [103] M. Solak, M. Kılıç, Y. Hüseyin and A. Şencan, "Removal of Suspended Solids and Turbidity from Marble Processing Wastewaters by Electrocoagulation: Comparison of Electrode Materials and Electrode Connection Systems," Journal of Hazardous Materials, Vol. 172, No. 1, 2009, pp. 345-352. doi:10.1016/j.jhazmat.2009.07.018

103. [104] W. Chou, C. Wang and K. Huang, "Investigation of Process Parameters for the Removal of Polyvinyl Alcohol from Aqueous Solution by Iron Electrocoagulation," Desalination, Vol. 251, No. 1-3, 2010, pp. 12-19. doi:10.1016/j.desal.2009.10.008

104. [105] W. Chou, C. Wang, K. Huang and T. Liu, "Electrochemical Removal of Salicylic Acid from Aqueous Solutions Using Aluminum Electrodes," Desalination, Vol. 271, No. 1-3, 2011, pp. 55-61. doi:10.1016/j.desal.2010.12.013

105. [106] M. Malakootian, H. J. Mansoorian and M. Moosazadeh, "Performance Evaluation of Electrocoagulation Process Using Iron-Rod Electrodes for Removing Hardness from Drinking Water," Desalination, Vol. 255, No. 1-3, 2010, pp. 67-71. doi:10.1016/j.desal.2010.01.015

106. [107] A. A. Bukhari, "Investigation of the Electro-Coagulation Treatment Process for the Removal of Total Suspended Solids and Turbidity from Municipal Wastewater," Bioresource Technology, Vol. 99, No. 5, 2008, pp. 914-921.doi:10.1016/j.biortech.2007.03.015

107. [108] M. A. Rodrigo, P. Cañizares, C. Buitrón and C. Sáez, "Electrochemical Technologies for the Regeneration of Urban Wastewaters," Electrochimica Acta, Vol. 55, No. 27 2010, pp. 8160-8164. doi:10.1016/j.electacta.2010.01.053

108. [109] K. Sadeddin, A. Naser and A. Firas, "Removal of Turbidity and Suspended Solids by Electro-Coagulation to Improve Feed Water Quality of Reverse Osmosis Plant," Desalination, Vol. 268, No. 1-3, 2011, pp. 204-207. doi:10.1016/j.desal.2010.10.027

109. [110] M. G. Kılıç, Ç. Hoşten and Ş. Demirci, "A Parametric Comparative Study of Electrocoagulation and Coagulation Using Ultrafine Quartz Suspensions," Journal of Hazardous Materials, Vol. 171, No. 1-3, 2009, pp. 247- 252.doi:10.1016/j.jhazmat.2009.05.133

110. [111] D. Ghernaout, B. Ghernaout and A. Boucherit, "Effect of pH on Electrocoagulation of Bentonite Suspensions in Batch Using Iron Electrodes," Journal of Dispersion Science and Technology, Vol. 29, No. 9, 2008, pp. 1272- 1275. doi:10.1080/01932690701857483

111. [112] S. Gao, J. Yang, J. Tian, F. Ma, G. Tu and M. Du, "Electro-Coagulation-Flotation Process for Algae Removal," Journal of Hazardous Materials, Vol. 177, No. 1-3, 2010, pp. 336-343. doi:10.1016/j.jhazmat.2009.12.037

112. [113] D. Ghernaout, A. Badis, A. Kellil and B. Ghernaout, "Application of Electrocoagulation in Escherichia coli Culture and Two Surface Waters," Desalination, Vol. 219, No. 1-3, 2008, pp. 118-125. doi:10.1016/j.desal.2007.05.010

113. [114] C. Ricordel, A. Darchen and D. Hadjiev, "Electrocoagulation-Electroflotation as a Surface Water Treatment for Industrial Uses," Separation and Purification Technology, Vol. 74, No. 3, 2010, pp. 342-347. doi:10.1016/j.seppur.2010.06.024

114. [115] D. Ghernaout, B. Ghernaout, A. Saiba, A. Boucherit and A. Kellil, "Removal of Humic Acids by Continuous Electromagnetic Treatment Followed by Electrocoagulation in Batch Using Aluminium Electrodes," Desalination, Vol. 239, No. 1-3, 2009, pp. 295-308.doi:10.1016/j.desal.2008.04.001

115. [116] M. Vepsäläinen, M. Ghiasvand, J. Selin, J. Pienimaa, E. Repo, M. Pulliainen and M. Sillanpää, "Investigations of the Effects of Temperature and Initial Sample pH on Natural Organic Matter (NOM) Removal with Electrocoagulation Using Response Surface Method (RSM)," Separation and Purification Technology, Vol. 69, No. 3, 2009, pp. 255-261.doi:10.1016/j.seppur.2009.08.001

116. [117] X. Li, Q. Feng, Q. Meng and Y. Ceng, "Electrocoagulation for the Drinking Water Treatment of Polluted Surface Water Supplies," 2nd International Conference on Bioinformatics and Biomedical Engineering, iCBBE 2008, Shanghai, May 2008, pp. 3091-3094.

117. [118] N. Uduman, V. Bourniquel and M. K. Danquah and A. F. A. Hoadley, "A Parametric Study of Electrocoagulation as a Recovery Process of Marine Microalgae for Biodiesel Production," Chemical Engineering Journal, Vol. 174, No. 1, 2011, pp. 249-257.doi:10.1016/j.cej.2011.09.012

118. [119] D. Vandamme, S. C. V. Pontes, K. Goiris, I. Foubert, L. J. J. Pinoy and K. Muylaert, "Evaluation of Electro-Coagulation-Flocculation for Harvesting Marine and Freshwater Microalgae," Biotechnology and Bioengineering, Vol. 108, No. 10, 2011, pp. 2320-2329.doi:10.1002/bit.23199

119. [120] O. Yahiaoui, L. Aizel, H. Lounici, N. Drouiche, M. F. A. Goosen, A. Pauss and N. Mameri, "Evaluating Removal of Metribuzin Pesticide from Contaminated Groundwater Using an Electrochemical Reactor Combined with Ultraviolet Oxidation," Desalination, Vol. 270, No. 1-3, 2011, pp. 84-89. doi:10.1016/j.desal.2010.11.025

120. [121] A. E. Yilmaz, R. Boncukcuoğlu, M. M. Kocakerim, M. T. Yilmaz and C. Paluluoğlu, "Boron Removal from Geothermal Waters by Electrocoagulation," Journal of Hazardous Materials, Vol. 153, No. 1-2, 2008, pp. 146-151. doi:10.1016/j.jhazmat.2007.08.030

121. [122] C. P. Nanseu-Njiki, S. R. Tchamango, P. C. Ngom, A. Darchen and E. Ngameni, "Mercury(II) Removal from Water by Electrocoagulation Using Aluminium and Iron Electrodes," Journal of Hazardous Materials, Vol. 168, No. 2-3, 2009, pp. 1430-1436.doi:10.1016/j.jhazmat.2009.03.042

122. [123] W. Wan, T. J. Pepping, T. Banerji, S. Chaudhari and D. E. Giammar, "Effects of Water Chemistry on Arsenic Removal from Drinking Water by Electrocoagulation," Water Research, Vol. 45, No. 1, 2011, pp. 384-392. doi:10.1016/j.watres.2010.08.016

123. [124] S. Vasudevan, J. Lakshmi and G. Sozhan, "Studies on the Removal of Iron from Drinking Water by Electrocoagulation—A Clean Process," Clean—Soil, Air, Water, Vol. 37, No. 1, 2009, pp. 45-51. doi:10.1002/clen.200800175

124. [125] D. Ghosh, C. R. Medhi and M. K. Purkait, "Treatment of Drinking Water Containing Iron Using Electrocoagulation," International Journal of Environmental Engineering, Vol. 2, No. 1-3, 2010, pp. 212-227.

125. [126] S. Vasudevan, S. M. Sheela, J. Lakshmi and G. Sozhan, "Optimization of the Process Parameters for the Removal of Boron from Drinking Water by Electrocoagulation—A Clean Technology," Journal of Chemical Technology and Biotechnology, Vol. 85, No. 7, 2010, pp. 926-933. doi:10.1002/jctb.2382

126. [127] S. Vasudevan, J. Lakshmi and G. Sozhan, "Effects of Alternating and Direct Current in Electrocoagulation Process on the Removal of Cadmium from Water," Journal of Hazardous Materials, Vol. 192, No. 1, 2011, pp. 26- 34.

127. [128] E. Keshmirizadeh, S. Yousefi and M. K. Rofouei, "An Investigation on the New Operational Parameter Effective in Cr(VI) Removal Efficiency: A Study on Electrocoagulation by Alternating Pulse Current," Journal of Hazardous Materials, Vol. 190, No. 1-3, 2011, pp. 119-124. doi:10.1016/j.jhazmat.2011.03.010

128. [129] M. S. Bhatti, D. Kapoor, R. K. Kalia, A. S. Reddy and A. K. Thukral, "RSM and ANN Modeling for Electrocoagulation of Copper from Simulated Wastewater: Multi Objective Optimization Using Genetic Algorithm Approach," Desalination, Vol. 274, No. 1-3, 2011, pp. 74-80. doi:10.1016/j.desal.2011.01.083

129. [130] M. Behbahani, M. R. A. Moghaddam and M. Arami, "Techno-Economical Evaluation of Fluoride Removal by Electrocoagulation Process: Optimization through Response Surface Methodology," Desalination, Vol. 271, No. 1-3, 2011, pp. 209-218.doi:10.1016/j.desal.2010.12.033

130. [131] W. Chou and Y. Huang, "Electrochemical Removal of Indium Ions from Aqueous Solution Using Iron Electrodes," Journal of Hazardous Materials, Vol. 172, No. 1, 2009, pp. 46-53. doi:10.1016/j.jhazmat.2009.06.119

131. [132] A. Shafaei, M. Rezayee, M. Arami and M. Nikazar, "Removal of Mn^{2+} Ions from Synthetic Wastewater by Electrocoagulation Process," Desalination, Vol. 260, No. 1-3, 2010, pp. 23-28. doi:10.1016/j.desal.2010.05.006

132. [133] E. Lacasa, P. Cañizares, C. Sáez, F. J. Fernández and M. A. Rodrigo, "Electrochemical Phosphates Removal Using Iron and Aluminium Electrodes," Chemical Engineering Journal, Vol. 172, No. 1, 2011, pp. 137-143. doi:10.1016/j.cej.2011.05.080

133. [134] S. Vasudevan, J. Lakshmi, J. Jayaraj and G. Sozhan, "Remediation of Phosphate-Contaminated Water by

Electrocoagulation with Aluminium, Aluminium Alloy and Mild Steel Anodes," Journal of Hazardous Materials, Vol. 164, No. 2-3, 2009, pp. 1480-1486.doi:10.1016/j.jhazmat.2008.09.076

134. [135] Z. V. P. Murthy and S. Parmar, "Removal of Strontium by Electrocoagulation Using Stainless Steel and Aluminum Electrodes," Desalination, Vol. 282, 2011, pp. 63-67. doi:10.1016/j.desal.2011.08.058

135. [136] G. Moussavi, F. Majidi and M. Farzadkia, "The Influence of Operational Parameters on Elimination of Cyanide from Wastewater Using the Electrocoagulation Process," Desalination, Vol. 280, No. 1-3, 2011, pp. 127-133. doi:10.1016/j. desal.2011.06.052

136. [137] M. Pociecha and D. Lestan, "Using Electrocoagulation for Metal and Chelant Separation from Washing Solution after EDTA Leaching of Pb, Zn and Cd Contaminated Soil," Journal of Hazardous Materials, Vol. 174, No. 1-3, 2010, pp. 670-678. doi:10.1016/j.jhazmat.2009.09.103

137. [138] K. Mansouri, K. Elsaid, A. Bedoui, N. Bensalah and A. Abdel-Wahab, "Application of Electrochemically Dissolved Iron in the Removal of Tannic Acid from Water," Chemical Engineering Journal, Vol. 172, No. 2-3, 2011, pp. 970-976.doi:10.1016/j. cej.2011.07.009

138. [139] N. Mameri, A. R. Yeddou, H. Lounici, D. Belhocine, H. Grib and B. Bariou, "Defluoridation of Septentrional Sahara Water of North Africa by Electrocoagulation Process Using Bipolar Aluminium Electrodes," Water Research, Vol. 32, No. 5, 1998, pp. 1604-1612. doi:10.1016/S0043-1354(97)00357-6

139. [140] M. Karhu, V. Kuokkanen, T. Kuokkanen and J. Rämö, "Bench Scale Electrocoagulation Studies of Bio Oil-inWater and Synthetic Oil-in-Water Emulsions," Separation and Purification Technology, Vol. 96, 2012, pp. 296- 305.doi:10.1016/j.seppur.2012.06.003

Effect of Chytriomyces Hyalinus on Industrial Wastewater Pre-treated with Electrocoagulations in a Continuous System

Moisés Tejocote-Pérez[1,2], Patricia Balderas-Hernández[1], Carlos Barrera-Díaz[1], Gabriela Roa-Morales[1], And Victor Bárcena[2]

[1]Sustainable Chemistry Research Center, State of Mexico Autonomy University and Mexico Autonomy National University (UAEM-UNAM), Toluca, Mexico

[2]Sciences Faculty, State of Mexico Autonomy University, Cerrillo Campus, Toluca, Mexico

ABSTRACT

A strain of Chytriomyces hyalinus fungus was applied as a pretreatment on industrial wastewater pollutant using electrocoagulations column of aluminum electrodes in a continuous system. The parameters considered in this experiment include pH, conductivity, color, turbidity,

COD (Chemical Oxygen Demand), BOD (Biochemical Oxygen Demand) nitrate, nitrite, and SB (sporangia biomass). Biological and electrocoagulations treatments had the next conditions: Chytriomyces hyalinus solutions 1:10, 60 min of biological treatment, 50 mL/min flow, constant ventilations, 15 min of electrocoagulations time and 3.4 A of electrical current. Color and turbidity values dropped with a 90% efficiency (2700 to170 Pt-Co; 120 to10 FAU, respectively), COD 68% (2100 to 672 mg/L), BOD_5 70% (650 to 195 mg/L), nitrate showed an 86% (3.8 to 0.5 mg/L), finally nitrite with a 60% amount reduction (1.5 to 0.6 mg/L). For SB parameter, there was a value rising as same as the treatment time ($r^2 = 0.90$) carrying out a $y = 94.302^{e0.0356x}$ model. These results reveal a positive outcome of Chytriomyces hyalinus on industrial wastewater pollutants pre-treated with aluminium electrocoagulations in a continuous system.

INTRODUCTION

Chytriomyces hyalinus, is a fresh water fungus which degrades organic matter as well as carbohydrates on adverse conditions of water quality [1,2]. It was unknown, until recently, how the proteollytic enzymes that allow the fungus survival in adverse conditions, remove the organic pollutants [3].

Some recordings about fungus aerobic treatments on industrial wastewater presents a 90% reduction in color, turbidity and phenols amount [4,5], an 80% in ammonia denitrification [6-8], 50% Cr (VI) loss [9] also 80 % nitrite and nitrate diminution [2]. Some fungi species such as Aspergillus Niger, Aspergillus oryzae, Penicillium corylophilum and Trichoderma viridae can be used as treatments with a 50% - 80% efficiency range showing a COD decrease on industrial wastewater [10-12]. Aspergillus oryzae and Rhizopus oligosporus FBP (Fungal Biomass Protein) remove in a 75% the ammonia amount on industrial wastewater with anaerobic treatments [10].

Based on these traits, Chytriomyces hyalinus can be employed as industrial wastewater biological treatment owing to the fact that degrades soluble and colloidal organic matter. Such treatment consists in the oxidation of the organic matter by bacterial, fungus, protozoa and microalgae consortiums [7,8,13-18]. In this procedure the microorganisms reduce the COD and BOD [19,20].

In many cases, industrial wastewater inhibit the degradation ability of the biological treatments probably due to its effluent character that commonly contains several organic and inorganic pollutants such as phenols, solvents, aromatics, organic matter, metals, dyes, nitrate, nitrite, chloride and salts [6,7,21-24]. In order to solve this problem and increase the efficiency, some authors coupled them with ozono, photocatalysis [4] filtrations [22], Fenton reactions [25,26], chemical coagulants [27, 28] and lastly Fe and Al electrocoagulantion [25,29,30]. Electrocoagulations consists in the formation of metallic hydroxide such as $Fe(OH)_2$ and $Al(OH)_3$ within wastewater by electrodissolution of a soluble anode, this method frequently uses electrodes made of iron (Fe) or aluminum (Al); four phases can be observed: firstly an electrolytic reaction at electrode surfaces, secondly the formation of coagulants in the aqueous phase, then the adsorption of soluble or colloidal pollutants on metallic coagulants, and finally the removal by sedimentation or flotation [25,31,32]. The reactions to aluminum electrodes are shown next:

$$\text{Anode: } Al \rightarrow Al(aq)^{3+} + 3e \tag{1}$$

$$\text{Cathode: } 3H_2O + 3e \rightarrow 3/2H_2 + 3OH^- \tag{2}$$

$$\text{Anode: } 2Al + 6H_2O + 2OH^- \rightarrow 2\,Al(OH)_4^- + 3H_2 \tag{3}$$

The $Al(aq)^{3+}$ and OH^- ions are generated by an anode (1) and a cathode (2) form several monomeric species previous to $Al(OH)_3$ coagulant (3) which oxides the pollutants and remove them by sedimentation [25], in this process the electrocoagulant $Al(OH)_3$ increases the wastewater pollutant bioavailability to the biological treatments [33].

Lately, the electrochemical treatment of wastewater applies a continuous systems since the procedure increases the wastewater pollutants bioavailability and operate with dynamic treatment plants parameters such as flows, ventilations, hydraulic and organic charge, liquor mix and resident times [14,19,20,34,35].

Due to the bioavailability increment of the industrial wastewater pollutants to biological treatments by electrocoagulations was noticed. The present work evaluates the effect of Chytriomyces hyalinus, whose has not been used as biological treatment for industrial effluents, on industrial wastewater pre-treated with aluminum electrodes in a continuous system.

MATERIALS AND METHODS

Wastewater Samples

The industrial wastewater was collected in an industrial treatment plant in México State; 10 L samples of water were taken prior to the primary clarifiers every month during March-November 2011. The samples were transported in 10 L plastic containers and stored to 4° for 24 hrs following the Mexican [36] and International [37] normative.

Chytriomyces hyalinus and Electrocoagulations Pretreatment in a Continuous System

The continuous system applied to evaluate the effects of Chytriomyces hyalinus on electrocoagulation pre-treated industrial wastewater, shown in Figure 1, are based on [35] and [33] systems. The continuous system contains an electrocoagulation column coupled with a Chytriomyces hyalinus aerobic biological reactor.

Electrocoagulations Reactor

The electrocoagulation reactor contains four circular aluminum electrodes arranged vertically into an 11×20 cm column, the electrodes area is 63.61 cm^2 and 0.05 A/cm^2 of current density. The system operates with 1 L of wastewater volume, a peristaltic pump, constant flow of 50 ml/min, 3.4 A of electrical current, 2.0 psi of ventilations, temperature of 19°C and 15 min of electrochemical pretreatment [29,33].

Biological Reactor with Chytriomyces Hyalinus

Chytriomyces hyalinus strain was obtained from isolations made to freshwater samples, which were taken from a water body near to Lerma River in Mexico State. Five Chytriomyces hyalinus isolations were made in aseptic conditions using queratine extract medium (KEM) containing 100 mL of queratine liquid extract, 0.5 g NaCl, 0.2 g MgCl, 0.05 g glucose and 1000 mL distilled water. The isolations were incubated up to seven days maintaining 20°C. The biomass of Chytriomyces hyalinus developed around organic matter flocks in the aseptic medium was removed by aseptic vacuum filtrations using filter paper Whatman 4, after that the filtering obtained was inoculated in the biological reactor, described in the Figure 1, with 1:1 concentrations of Chytriomyces hyalinus strains and sterile distilled water, this activated sludge was maintained with 2.0 psi of ventilation during 8 hrs previous to electrocoagulation pre-treatment [24,38].

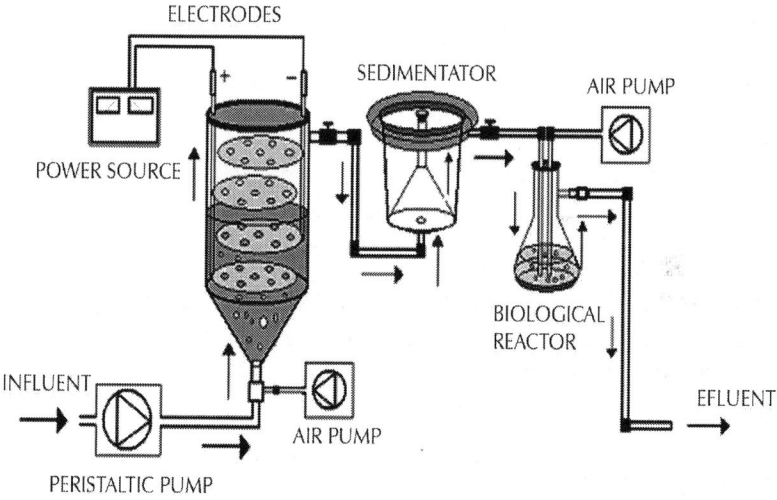

Figure 1: Electrocoagulations-Chytriomyces hyalinus continuous system.

The electrocoagulation wastewater pre-treatment was conducted to Chytriomyces hyalinus biological reactor by peristaltic pump, Figure 1. The biological reactor was operated with 2.0 psi of ventilations, temperature of 19°C, and constant flow of 50 ml/min and 60 min of biological treatment.

The Chytriomyces hyalinus strain was characterized utilizing an optical microscopic considering the morphology and sporangia biomass (SB) using KOH 5%, congo red as biological dye and a Leica microscope GME model, the SB in activated sludge and liquor mix was 345 sporangia/mL, this obtained using a Marienfeld Neubauer chamber [39]. The industrial pollutants toxic effects in the biological treatment was evaluated with the SB (sporangia number/mL of liquor mix) during the biological treatment as well [24,39].

Wastewater Characterizations

Wastewater was characterizes using pH, conductivity, color, turbidity, BOD_5, COD, nitrate and nitrite parameters according to the Mexican [36,40,41] and International [37] normative.

PH and Conductivity

Conductivity and pH were measured directly on wastewater before and after electrochemical and biological treatments using an OAKTON potentiometer 35631-60 model with standard calibration solutions [37, 42].

Color and Turbidity

The color and turbidity was evaluated directly on wastewater previous and subsequent to electrochemical and biological treatments applying the standard scale color method of platinum-cobalt (Pt-Co) with a range from 0 to 500 units Pt-Co and the turbidity FAU units' method with a range from 0 to 5000 units. The reads from both methods were made using a Hach spectrophotometer DR/4000U model with 465 and 860 nm wavelength respectively [37, 42].

COD and BOD$_5$

The COD parameter was used to evaluate the amount of organic and inorganic pollutants oxidized by chemical digest with chromic and sulfuric acids in potassium dichromate solutions, the digest was regulated in a thermoreactor Thermo Elec Corp COD Orion model from 125°C to 150°C, the reaction was maintained during 2hrs, eventually the evaluations were expressed in mg/L units by a Hach spectrophotometer DR/4000U model, The BOD_5 test was employed to evaluate the oxygen amount needed by microorganisms to oxidize the organic matter during five days of incubation to 20°C, this test was made using BOD_5 Hach bottles kit, a dissolved oxygen meter OAKTON DO 110 model and a incubator VWR Scientific 1535 model [37,42].

Nitrate and Nitrite

The nitrates and nitrites were quantified before and after the electrochemical and biological treatments, such quantifications were made using the Nitraver® X and Nitriver® 3 methods of Hach besides the values were determinate by a spectrophotometer Hach DR/4000U. These quantifications were made according to the normative [37, 42].

UV-Visible Spectrophotometer

Some water samples previous and behind electrochemical-biological treatments were analyzed by spectrophotometer UV-visible Perkin Elmer Lambda 25 model (200 - 900 nm) this technique is necessary to identify the behavior of some organic pollutants considering their absorption properties [32].

RESULTS AND DISCUSSIONS

Parameters

The pH, conductivity, color, turbidity, COD, BOD_5, nitrate and nitrite values of industrial wastewater before and after continuous coupled systems are shown in Table 1.

PH and Conductivity

The pH value intensified from 7.45 ± 0.84 to 8.45 ± 0.50 in consequence of continuous system, this behavior is shown in Figure 2(a). The pH boosted as a result of the OH⁻ radicals production on cathode surfaces from aluminum electrodes into the electrochemical column and the OH⁻ radicals electrodisolution in the aluminum hydroxide $Al(OH)_3$ reactions. The pH values before and after the treatment corresponding to common industrial wastewater with 7.0 and 9.0 values is a beneficial condition to be applied in the electrochemical and biological treatments [33].

The Figure 2(b) divulge the conductivity abate form 6.7 ± 0.44 to 4.57 ± 0.30 mS, this tendency is caused by Cl⁻ as well as chlorides salts consumption in electrocoagularion redox reactions [25,29,32].

The conductivity values indicate a beneficial condition applying only 15 min electrochemical pulse in the coupled treatment. Furthermore there is a gain in the pollutants bioavailability to Chytriomyces hyalinus [14,19, 20,33-35].

Color and Turbidity

The color displays high values in wastewater samples earlier from the coupled treatment; additionally it was produced by dissolved organic matter from chemical and food industries effluents [29]. Literature points Aspergillus and Penicillium as color discharge up to 90% in coupled systems, for these experiment Figure 2(c) shows a 93% color removal using Chytriomyces hyalinus, a 50% more efficient than conventional activated sludge [4, 5, 10, 30, and 32]. It is to say that electrochemical pretreatment enhance the industrial pollutant bioavailability to Chytriomyces hyalinus.

As perceived in Figure 2(c), the turbidity had a similar behavior pointing that the values diminish from 120 to 10 FAU with 91% efficiency after using Chytriomyces hyalinus electrocoagulation system. Thus only with Chytriomyces hyalinus the efficiency rate was higher than 60% reported to conventional biological treatment for activated sludge [4, 5, 10, 30, 32].

Both, color and turbidity parameters register a shrinks after a 15-min of biological treatment, meaning that the pollutant bioavailability improves in coupled condition of continuous system.

Table 1: Values of pH, conductivity, color, turbidity, COD, BOD$_5$, nitrate and nitrites before and after the electrocoagulations-Chytriomyces hyalinus treatment

	pH	Conductivity (mS)	Color (Pt-Co)	Turbidity (FAU)	COD (mg/L)	BOD5 (m8/14	Nitrate (mg/L)	Nitrite (mg/L)
Before	7.45 ± 0.84	6.7 ± 0.44	2700 ± 51.43	120 ± 19.30	2100 ± 42.86	650 ± 23.12	3.8 ± 0.41	1.5 ± 0.22
After	8.45 ± 0.50	4.57 ± 0.30	170 ± 2.0	10 ± 1.2	672± 10.6	195 ± 4.1	0.5 ± 0.20	0.6 ± 0.10
%			93	91	68	70	86	60

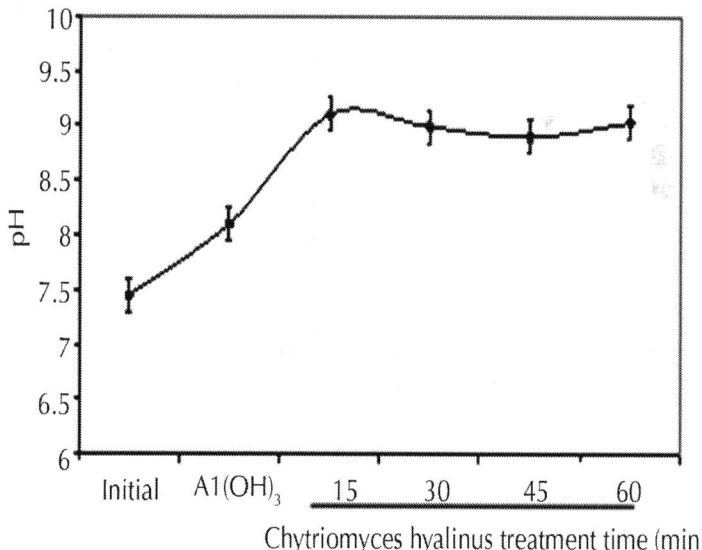

Chytriomyces hyalinus treatment time (min)

(a)

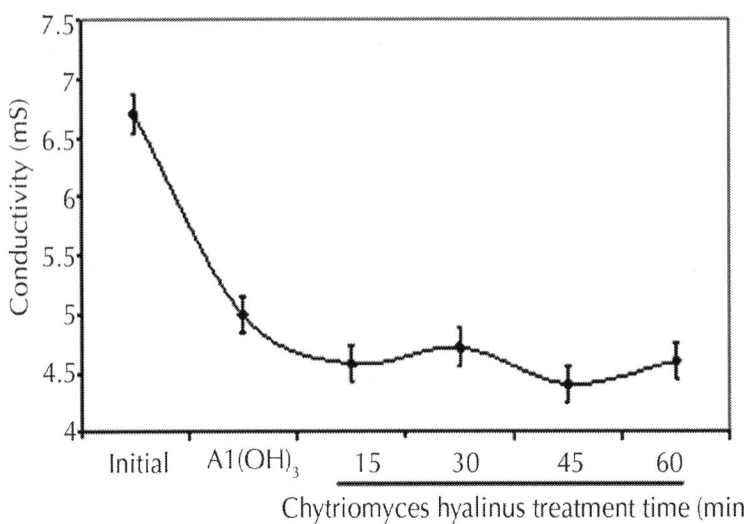

(b)

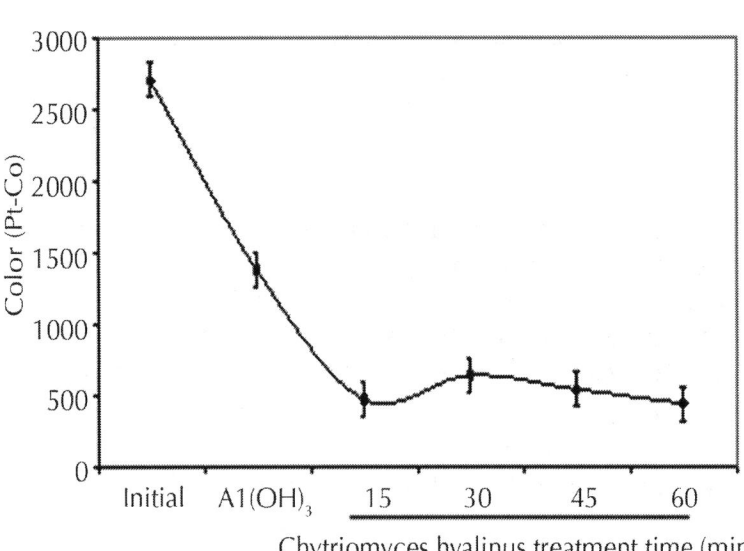

(c)

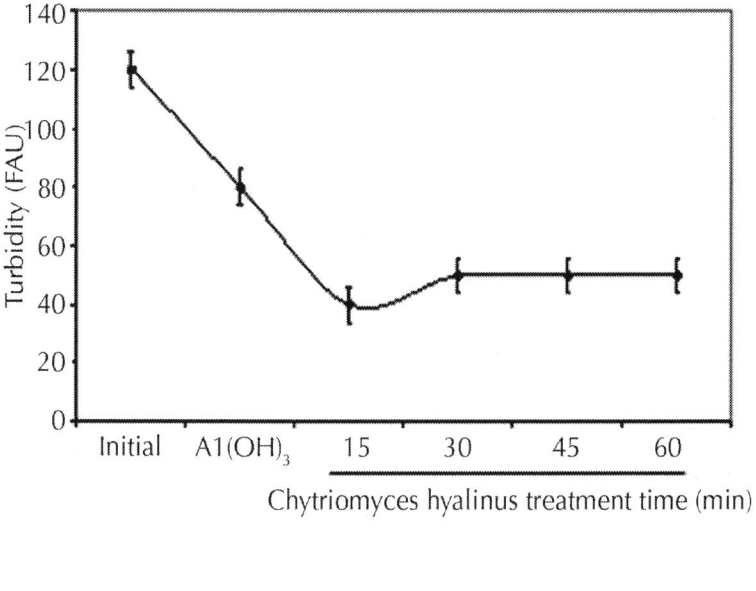

(d)

Figure 2: (a) pH; (b) Conductivity; (c) Color and (d) Turbidity during the electrochemical-Chytriomyces hyalinus treatment.

COD and BOD$_5$

The COD and BOD$_5$ values in Figure 3, show a pollutant removal efficiency declining, 68% and 70%, respectively [8,11,12]. Such fact indicates that Aspergillus niger, Aspergillus oryzae, Penicillium corylophilum and Trichoderma viridae tends to contract in a 50% COD and BOD$_5$ levels, exposing a greater reaction rather than the one presented by reference [8,11,12].

The COD and BOD$_5$ results reveal three important points. In first place, Aspergillus, Penicillium and Trichoderma strains share a pluricellular organization while Chytriomyces hyalinus posses an unicellular level, such biological advantage rises the biomass presence like the resident time during the treatment, whereas this datum represent an inconvenience for Chytriomyces hyalinus. Moreover the COD and BOD$_5$ upshots presents the same efficiency rate; even using a lower resident time [8,11, 12].

In second place, COD and BOD_5 parameters can now be considered as a new contribution to biological treatment of activated sludge since the values were lessen by using fungus strains only, while the conventional treatments of activated sludge use several microorganisms consortiums [7,8,13-18].

In third place, it stresses the coupled continuous system role in the removal efficiency owing to the electrocoagulation conditions that raises the bioavailability of industrial wastewater pollutants to Chytriomyces hyalinus. [14, 19, 20, 25, 29, 33-35].

Making this study the first report of industrial pretreated wastewater by electrocoagulation coupled with Chytriomyces hyalinus as a biological system.

Nitrate and Nitrite

Nitrates (NO_3^-) and nitrites (NO_2^-) dwindled subsequent to the biological treatment from 3.8 to 0.5 mg/L and from 1.5 to 0.6 mg/L with an effectiveness of 86% and 60% respectively. Such behavior is displayed in Figure 3.

Nitrates were more affected than nitrites because its chemical nitrogen trait, having as a result a higher biological assimilation thanks to aquatic microorganisms [2,6-8].

These nitrogenous compounds, industrial pollutants type, are presented naturally in the experimental effluent, therefore it is important to make a continuous revision

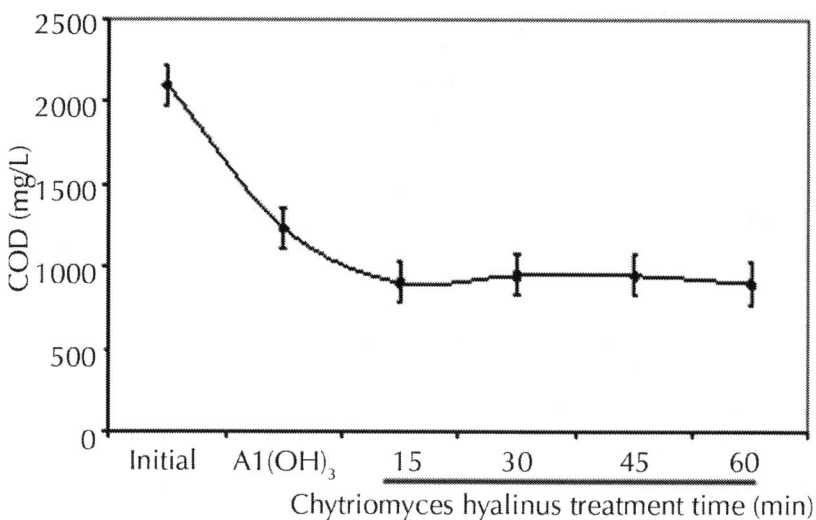

(a)

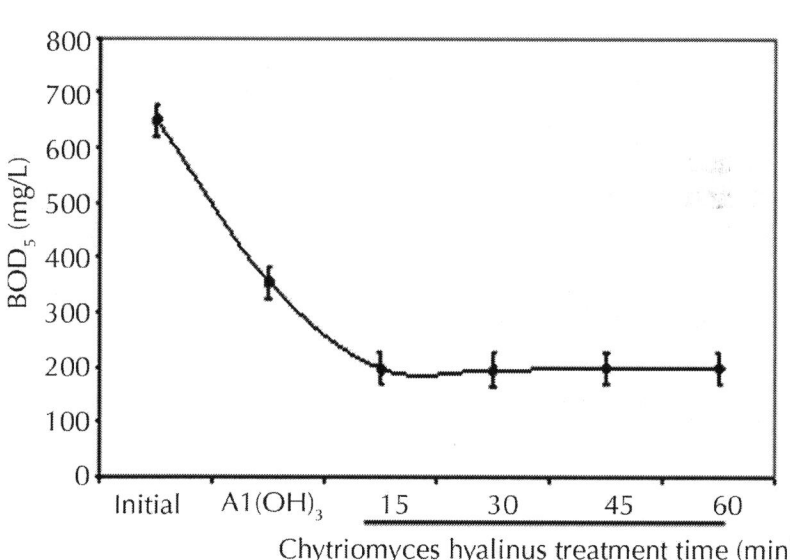

(b)

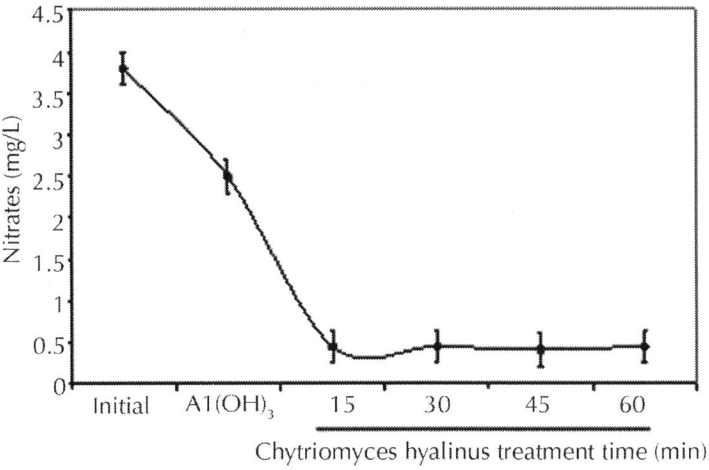

(c)

(d)

Figure 3: (a) COD, (b) BOD$_5$, (c) nitrates and (d) nitrites during the electro-chemical-Chytriomyces hyalinus treatment.

During the coupled treatment because the pollutant removal tendency in the continuous system can manifest. The nitrogenous compounds removal grew when they were oxidize, nitrates into nitrites by electrocoagulation treatments [25,31,32]. The nitrates amount and removal efficiency were higher than nitrites due to the reduction from nitrites to nitrates at the electrochemical reactor besides the quickly assimilation by Chytriomyces hyalinus in the biological reactor.

This reduction is an evidence of Chytriomyces hyalinus active metabolism in addition to water denitrification in aerobic conditions as its natural ability [2-4].

Similar industrial pollutant denitrification greater than 50% were reported by [6-8,10] with Aspergillus oryzae and Rhizopus oligosporus.

Sporangia Biomass (SB)

Figure 4 shows Chytriomyces hyalinus SB concentrations on a 60-min session of biological treatment. The biomass suffered an increment as a result of 30 min by the time sporangia amount increased. The SB tendency indicates that Chytriomyces hyalinus is capable of resisting the pollutant conditions in the liquor mix, noticed by an exponential growth $y = 94.302^{e^{0.0356x}}$ [10, 24, 39].

SB value ascended when COD and BOD_5 values decrease, this reaction denotes that pollutants bioavailability to Chytriomyces hyalinus were modified within a 15- min of electrochemical treatment, hence the toxic effect of industrial wastewater on biological treatment was reduced; opposite to the frequent problem in bio-logical treatment inhibitions [6,7,21-24].

Samanthi and Chandralata (2009) report an optimal growth for Chytriomyces hyalinus in some aquatic systems, showing pH values from 6.8 to 8.5, electrochemical and biological conditions in the treatment were kept into these pH values, consequently the pollutants removal suffered an increment.

The sludge amount after biological treatments with Chytriomyces hyalinus was 2 g/L, within normal range from 0.5 to 5 g/L was reported in laboratory level experiments [21-24].

UV-Visible Spectrophotometer Characterizations

A pollutant decrement followed by electrocoagolutions Chytriomyces hyalinus system treatment can be observed in Figure 5. The absorbance indicates a spectral reduction with a 60% efficiency, showing an absorbance result of 400, 475 and 625 nm corresponding to phenols, solvents, aromatic and organic matter; similar wastewater spectra cases have been reported [29,32,33]. The current tendency was consistent with COD and BOD_5 results.

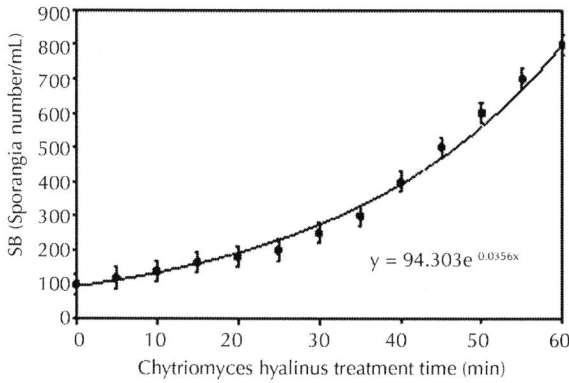

Figure 4: Sporangia biomass and Chytridium hyalinus treatment time.

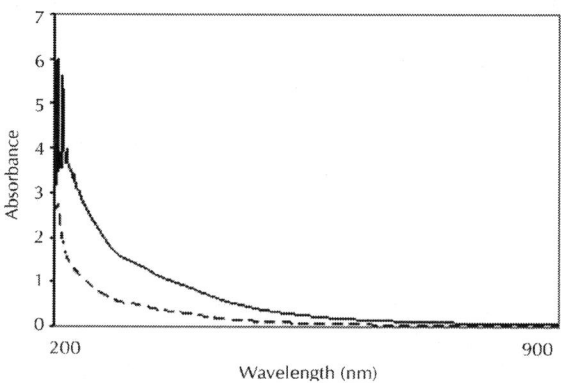

Figure 5: UV visible characterizations of industrial wastewater before (-) and after (----) electrocoagulations-Chytriomyces hyalinus treatment.

CONCLUSIONS

All in all the Chytryomyces hyalinus effect on industrial pre-treated wastewater by electrocoagulations in a continuous system had a positive efficiency on pollutants removal. The electrochemical pulse with aluminum electrodes extends the pollutants bioavailability to Chytryomyces hyalinus. Color and turbidity exhibited a reduction with 90% efficiency, COD 62%, BOD_5 69%, nitrate 86% and nitrite 60%. Chytryomyces hyalinus sporangial bio-mass (SB) heightens exponentially attending to a $y = 94.302e^{0.0356x}$ model, additionally increases when pollutant concentration fall, so as COD, BOD_5, nitrate and nitrite values. Finally the pollutants removed exposes a UV-visible spectra corresponding to organic pollutants.

ACKNOWLEDGEMENTS

The authors wish to acknowledge CONACYT and Universidad Autónoma Del Estado de México for the support given to this project 2794/2010-2011.

REFERENCES

1. S. Golubic, G. Radtke and T. Le, "Campion-Alsumard, Endolithic Fungi in Marine Ecosystems," Trends in Microbiology, Vol. 13, No. 5, 2005, pp. 229-234.doi:10.1016/j.tim.2005.03.007

2. J. Sumathi and R. Chandralata, "Anaerobic Denitrifications in Fungi from the Coastal Marine Sediments off Goa, India," Mycological Research, Vol. 113, No. 1, 2009, pp. 100-109. doi:10.1016/j.mycres.2008.08.009

3. G. Hageskal, N. Lima and I. Skaar, "The Study of Fungi in Drinking Water," Mycological Research, Vol. 113, No. 2, 2009, pp. 165-172. doi:10.1016/j.mycres.2008.10.002

4. M. Barreto-Rodríguez, J. V. Souza, S. E. Silva, T. F. Silva and C. B. T. Pavia, "Combined Photocatalytic and Fungal Processes for the Treatment of Nitrocellulose Industry Wastewater," Journal of Hazardous Materials, Vol. 161, No. 2-3, 2009, pp. 1569-1573. doi:10.1016/j.jhazmat.2008.05.012

5. T. Dalsgaard, D. E. Canfield, J. Petersen, B. Thamdrup and J. Acuña-González, "N2 Productions by the Anammox Reactions in the Anoxic Water Column of Golfo Dulce, Costa Rica," Nature, 422, No. 6932, 2003, pp. 606-608. doi:10.1038/nature01526

6. G. Buttiglieri, F. Malpei, E. Daverio, M. Melchiori, H. Nieman and J. Ligthart, "Denitrification of Drinking Water Sources by Advanced Biological Treatment Using a Membrane Bioreactor," Desalination, Vol. 178, No. 1-3, 2005, pp. 211-218. doi:10.1016/j.desal.2004.11.038

7. Y. T. Ahn, S. T. Kang, S. R. Chae, C. Y. Lee, B. U. Bae and H. S. Shin, "Simultaneous High-Strength Organic and Nitrogen Removal with Combined Anaerobic Upflow Bed Filter and Aerobic Membrane Bioreactor," Desalination, Vol. 202, No. 1-3, 2007, pp. 114-121. doi:10.1016/j.desal.2005.12.046

8. T. R. Thomsen, Y. Kong and P. H. Nielsen, "Ecophysiology of Abundant Denitrifying Bacteria In-Activated Sludge," Microbiology Ecological, Vol. 60, No. 3, 2007, pp. 370-382. doi:10.1111/j.1574-6941.2007.00309.x

9. K. Rajender, R. N. Bishnoi and K. G. Bishnoi, "Biosorption of Chromium Cr (VI) from Aqueous Solutions and Electroplating Wastewater Using Fungal Biomass," Chemical Engineering Journal, Vol. 135, No. 3, 2008, pp. 202- 209.

10. J. Bo, X. Q. Yan, Q. Yu and J. H. Van Leeuwen, "A Comprehensive Pilot Plant System for Fungal Biomass Protein Productions and Wastewater Reclamation," Advances in Environmental Research, Vol. 6, No. 2, 2002, pp. 179-189. doi:10.1016/S1093-0191(01)00049-1

11. A. Zahangir and A. Fakhru'l-Razi, "Enhanced Settlebility and Dewaterability of Fungal Treated Do-Mestic Wastewater Sludge by Liquid State Bioconversion Process," Water Research, Vol. 37, No. 5, 2003, pp. 1118-1124. doi:10.1016/S0043-1354(02)00452-9

12. Y. Z. Zhang, B. Jin, H. Z. Bai and Y. X. Wang, "Production of Fungal Biomass Protein Using Microfungi from Winery Wastewater Treatment," Bioresource Technology, Vol. 99, No. 9, 2008, pp. 3871-3876. doi:10.1016/j.biortech.2006.10.047

13. E. Liwarska-Bizukojc, "Application of Image Analysis Techniques in Activated Sludge Wastewater Treatment Processes,"

Biotechnology Letters, Vol. 27, No. 19, 2005, pp. 1427-1433. doi:10.1007/s10529-005-1303-2

14. D. K. Sharma, H. S. Saini, M. Singh, S. S. Chimni and B. S. Chadha, "Biological Treatment of Textile Dye Acid Violet-17 by Bacterial Consortium in an Up-Flow Immobilized Cell Bioreactor," Letters in Applied Microbiology, Vol. 38, No. 5, 2004, pp. 345-350. doi:10.1111/j.1472-765X.2004.01500.x

15. C. J. Van der Gast, A. S. Whiteley and I. P. Thompson, "Temporal Dynamics and Degradation Activity of a Bacterial Inoculum for Treating Waste Metal-Working Fluid," Environmental Microbiology, Vol. 6, No. 3, 2004, pp. 254-263. doi:10.1111/j.1462-2920.2004.00566.x

16. C. Kragelund, C. Levantesi, A. Borger, K. Thelen, D. Eikelboom, V. Tandoi, Y. Kong, J. Van der Waarde, J. Krooneman, S. Rossetti and T. R. ThomsenNielsen, "Identity, Abundance and Ecophysiology of Filamentous Chloroflexi Species Present in Activated Sludge Treatment Plants," Microbiological Ecology, Vol. 59, No. 3, 2007, pp. 671-682.doi:10.1111/j.1574-6941.2006.00251.x

17. S. Rossetti, M. Tomei, P. Nielsen and V. Tandoi, "Microthrix parvicella, a Filamentous Bacterium Causing Bulking and Foaming in Activated Sludge Sys-Tems: A Review of Current Knowledge," Microbiology Review, Vol. 29, No. 1, 2005, pp. 49-64.

18. V. L. Barbosa, S. D. Atkins, V. P. Barbosa, J. E. BurGess and R. M. Stuetz, "Characterization of Thiobacillus thioparus Isolated from an Activated Sludge Bioreactor Used for Hydrogen Sulfide Treatment," Journal of Applied Microbiology, Vol. 101, No. 6, 2006, pp. 1269-1281. doi:10.1111/j.1365-2672.2006.03032.x

19. F. Fatone, D. Bolzonella, P. Battistoni and F. Cecchi, "Removal of Nutrients and Micropollutants Treating Low Loaded Wastewaters in a Membrane Bioreactor Operating the Automatic Alternate-Cycles Process," Desalination, Vol. 183, No. 1-3, 2005, pp. 395-405.doi:10.1016/j.desal.2005.02.055

20. J. D. Jang, J. P. Barford, A. Lindawati and R. Renneberg, "Application of Biochemical Oxygen Demand (BOD) Biosensor for Optimization of Biological Carbon and Nitrogen Removal from Synthetic Wastewater in a Sequencing Batch Reactor System," Biosensor and Bioelectrode, Vol. 19, No. 8, 2004, pp. 805-812. doi:10.1016/j.bios.2003.08.009

21.   C. Della Rocca, V. Belgiorno and S. Meric, "Over-View of In-Situ Applicable Nitrate Removal Processes," Desalination, Vol. 204, No. 1-3, 2007, pp. 46-62.doi:10.1016/j.desal.2006.04.023

22.   Y. H. Kim, E. D. Hwang, W. S. Shin, J. H. Choi, T. W. Ha and S. J. Choi, "Treatments of Stainless Steel Wastewater Containing a High Concentration of Nitrate Using Reverse Osmosis and Nanomembranes," Desalination, Vol. 202, No. 1-3, 2007, pp. 286-292.doi:10.1016/j.desal.2005.12.066

23.   M. Milovanovic, "Water Quality Assesment and DeTermination of Pollution Sources along the Axios/Vardar River, Southeastern Europe," Desalination, Vol. 213, No. 1-3, 2007, pp. 159-173. doi:10.1016/j.desal.2006.06.022

24.   M. Ricart, E. Guasch, M. Alberch, D. Berceló, C. Bonnineau, A. Geiszinger, M. La Farré, J. Ferrer, F. Ricciardi, A. Romaní, S. Morín, L. Proia, L. Sala, D. Sureda and S. Sabater, "Triclosan Persistence through Wastewater Treated Plants and Its Potential Toxic Effects on Rever Biofilms," Aquatic Toxicology, Vol. 100, No. 4, 2010, pp. 346-353.doi:10.1016/j.aquatox.2010.08.010

25.   P. Cañizares, R. Paz, C. Sáez and M. A. Rodrigo, "Cost of the Electrochemical Oxidation of Wastewaters: A Comparison with Ozonations and Fenton Oxidation Processes," Journal of Environment Management, Vol. 90, No. 1, 2009, pp. 410-420. doi:10.1016/j.jenvman.2007.10.010

26.   K. V. Padoley, S. N. Mudliar, S. K. Banerjee, S. C. Deshmukh and R. A. Pandey, "Fenton Oxidation: A Pretreatment Option for Improved Biological Treatment of Pyridine and 3-Cyanopyridine Plant Wastewater," Chemical Engineering Journal, Vol. 166, No. 1, 2011, pp. 1-9. doi:10.1016/j.cej.2010.06.041

27.   O. Amuda and I. Amoo, "Coagulation/Flocculation Process and Sludge 5 Conditioning in Beverage Industrial Wastewater Treatment," Journal of Hazardous Materials, Vol. 141, No. 3, 2007, pp. 778-783. doi:10.1016/j.jhazmat.2006.07.044

28.   R. Braz, A. Pirra, M. Lucas and J. Peres, "Combinations of Long Term Aerated Storage and Chemical Coagulations/Flocculations to Winery Wastewater Treatment," Desalinations, Vol. 263, No. 1-3, 2010, pp. 226-232. doi:10.1016/j.desal.2010.06.063

29.   C. Barrera-Díaz, G. Roa-Morales, L. Avila-Cordoba, T. Pavon-Silva and B. Bilyeu, "Electrochemical Treatment Applied to

Food-Processing Wastewater Treatment," Industrial Engineering Chemical Research, Vol. 45, No. 1, 2006, pp. 34-38.doi:10.1021/ie050594k

30. M. Panizza and G. Cerisola, "Elechtrochemical Oxidation as a Final Treatment of Synthetic Tannery Wastewater," Environment Science Technology, Vol. 38, No. 20, 2004, pp. 5470-5475. doi:10.1021/es049730n

31. D. Rajkumar and K. Palanivelu, "Electrochemical Degradations of Cresols for Wastewater Treatment," Industrial Engineering Chemical Research, Vol. 42, No. 9, 2003, pp. 1833-1839. doi:10.1021/ie020759e

32. G. Roa-Morales, E. Campos-Medina, E. Aguilera-Cotero, B. Bilyeu and C. Barrera, "Aluminium Electrocuagulation with Peroxide Applied to Wastewater from Pasta and Cookie Processing," Separations and Purifications Technology, Vol. 54, No. 1, 2006. pp. 124-129. doi:10.1016/j.seppur.2006.08.025

33. M. Tejocote-Pérez, P. Balderas-Hernández, C. E BarreraDíaz, G. Morales and R. Natividad-Rangel, "Treatment of Industrial Effluents by a Continuous System: Electrocoagulation-Activated Sludge," Bioresource Technology, Vol. 101, No. 20, 2010, pp. 7761-7766.doi:10.1016/j.biortech.2010.05.027

34. D. W. Graham and V. H. Smith, "Designed Ecosystem Services: Application of Ecological Principles in Wastewater Treatment Engineering," Fronts in Ecology and the Environment, Vol. 2, No. 4, 2004, pp. 199-206. doi:10.1890/1540-9295(2004)002[0199:DESAOE]2.0.CO;2

35. N. N. Sang, S. Soda, K. Sei and M. Ike, "Effect of Aeration on Stabilization of Organic Solid Waste and Microbial Populations Dynamics in Lab-Scale Landfill Bioreactors," Journal of Bioscience and Bioengineering, Vol. 106, No. 5, 2008, pp. 425-432.doi:10.1263/jbb.106.425

36. NMX-AA-028-SCFI, "Water Analysis-Determination for Biochemical Oxygen Demand (BOD_5) in Natural, Wastewaters and Wastewaters Treated-Test Method," Diario Oficial de la Federación, México, 2005.

37. American Public Health Association, American Water Works Association, Water Environment Federation, "Standard Methods for the Examination of Water and Wastewater," Washington DC, Denver, Alexandria, 2005.

38. T. T. More, S. Yan, R. D. Tyagi and R. Y. Surampalli, "Potential Use of Filamentous Fungi for Wastewater Sludge Treatment," Bioresource Technology, Vol. 101, No. 20, 2010, pp. 7691-7700. doi:10.1016/j.biortech.2010.05.033

39. M. V. Garcia, A. C. Monteiro, M. J. P. Szabo, N. Prette and G. H. Bechara, "Mechanism of Infection and Colonization of Rhiphicephalus Sanguineus Eggs by Metarhizium anisopliae as Revealed by Scanning Electron Microscopy and Histopathology," Brazilian Journal of Microbiology, Vol. 36, No. 3, 2005, pp. 368-372. doi:10.1590/S1517-83822005000400012

40. NMX-AA-030-SCFI, "Water Analysis-Determination for Chemical Oxygen Demand (COD) in Natural, Wastewaters and Wastewaters Treated-Test Method," Diario Oficial de la Federación, México, 2001.

41. NMX-AA-034-SCFI, "Water Analysis-Determination of Salts and Solids Dissolved in Natural, Wastewaters and Wastewaters Treated-Test Method," Diario Oficial de la Federación, México, 2001.

42. HACH, "Water Analysis Manual," HACH Co., Loveland, 2008.

3

Photo-Fenton and Fenton Oxidation of Recalcitrant Industrial Wastewater Using Nanoscale Zero-Valent Iron

Henrik Hansson[1], Fabio Kaczala[1], Marcia Marques[1,2], and William Hogland[1]

[1]School of Natural Sciences, Linnaeus University (LNU), Landgången 3, 392 82 Kalmar, Sweden

[2]Department of Sanitary and Environmental Engineering, Rio de Janeiro State University (UERJ), São Francisco Xavier 524, 20551-013 Rio de Janeiro, RJ, Brazil

ABSTRACT

There is a need for the development of on-site wastewater treatment technologies suitable for "dry-process industries," such as the wood-floor sector. Due to the nature of their activities, these industries generate lower volumes of highly polluted wastewaters after cleaning activities. Advanced oxidation processes such as Fenton and photo-Fenton, are potentially feasible options for treatment of these wastewaters. One of the disadvantages of the Fenton process is the formation of large amounts of ferrous iron sludge, a constraint that might be overcome with the use of nanoscale zero-valent iron (nZVI) powder. Wastewater from a wood-floor industry with initial COD of 4956 mg/L and TOC of 2730 mg/L was treated with dark-Fenton (nZVI/H_2O_2) and photo-Fenton (nZVI/H_2O_2/UV) applying a 2-level full-factorial experimental design. The highest removal of COD and TOC (80% and 60%, resp.) was achieved using photo-Fenton. The supply of the reactants in more than one dose during the reaction time had significant and positive effects on the treatment efficiency. According to the results, Fenton and mostly photo-Fenton are promising treatment options for these highly recalcitrant wastewaters. Future investigations should focus on optimizing treatment processes and assessing toxic effects that residual pollutants and the nZVI might have. The feasibility of combining advanced oxidation processes with biological treatment is also recommended.

INTRODUCTION

The discharge of industrial wastewaters into either municipal sewerage system or directly into recipient water bodies has raised serious concerns during decades, leading to intensive research and development of on-site treatment technologies for industrial wastewater. However, whereas investigations have been focusing on industrial sectors that have water as an important input to their manufacturing processes, "dry-process industries" such as wood-floor and wood-furniture industries that have no water requirement in their production processes have been neglected [1, 2]. These industries generate wastewaters during cleaning and washing of machinery, surfaces, and floors and regardless of their relatively low volumes these cleaning wastewaters have very high chemical oxygen demand (COD) that varies from 3200 to 50,000 mg L^{-1} and the presence of recalcitrant organic compounds is a limiting factor for biological treatment in conventional centralized treatment plants. Dilution of 50 times or more with drinking water has been a common practice before discharging these wastewaters into the sewage system, which is not a sustainable strategy for the 21st century. The treatment of wastewaters from the timber industry using chemical methods has shown limited efficiency [3]. The use of biological treatment [1] and sorption/filtration processes [2] has also shown limitations when the main purpose is to comply with established standards for discharges into recipient water bodies. Advanced oxidation processes (AOPs) have previously been used to treat complex wastewaters in combinations with biological and/or chemical treatment. These studies have used AOP either after biological treatments to handle the most recalcitrant substances [4, 5] or before with the purpose of reducing toxicity, when the wastewater is too toxic for biological treatment [6, 7]. The Fenton reaction was first observed by H. J. Fenton in 1894 and is described as the enhanced oxidative power of H_2O_2 when using iron (Fe) as a catalyst under acidic conditions. It was later found that this enhancement was due to the generation of hydroxyl radical ($HO^•$) [8] that is one of the strongest oxidants ($E = 2.73$ V), and it is nonselective and capable of quickly oxidizing a broad range of organic pollutants [9]. AOPs are based on the generation of these very reactive species, which are the main oxidizing species in the Fenton process [10]. The Fenton's reagent alone or in combination has proven to be an effective way to degrade organic pollutants [11–13] and it has been used for the treatment of a wide variety of industrial wastewaters [14–16]. The Fenton process is a relatively economical method since it requires no additional energy when compared to many other AOPs. Furthermore, both iron and hydrogen peroxide are relatively cheap and safe. In the Fenton process, there is no mass transfer limitation, except during coagulation when high dose of the activator ferrous salt is needed [17]. The Fenton process is well known [18]. Iron (Fe) will in aqueous solution under acidic conditions oxidize to Fe^{2+} [19], which initiates the Fenton reaction:

$$2Fe^0 + 2H_2O \longrightarrow Fe^{2+} + H_2 + 2OH^- \tag{1}$$

The Fenton reaction is characterized by the catalytic decomposition of H_2O_2 as described below [20]:

$$Fe^{2+} + H_2O_2 \longrightarrow Fe^{3+} + OH^- + OH^• \tag{2}$$

The hydroxyl radicals created by the process can be scavenged by excess Fe^{2+} [20]:

$$OH^• + Fe^{2+} \longrightarrow OH^- + Fe^{3+} \tag{3}$$

This conventional Fenton process may be positively assisted by the application of UV-light [20]:

$$Fe^{3+} + H_2O + h\nu \longrightarrow OH^{\bullet} + Fe^{2+} + H^+ \qquad (4)$$

Photo-Fenton oxidation in the presence of short UV light (UV-C, 180–290 nm) [21] gives a faster oxidation as a consequence of the higher quantum yields [12]. When applying UV-C light, H_2O_2 can also be hydrolysed contributing to the $HO^{\bullet}$ formation [8]. The disadvantages of the Fenton process include (i) the formation of a high concentration of anions in the treated wastewater and (ii) large amounts of ferrous iron sludge [22]. Recent studies have attempted to overcome these drawbacks by applying nZVI together with H_2O_2 for industrial wastewater treatment [19, 23]. This alternative process could overcome the disadvantages associated with Fe^{2+}-based AOPs by using the solid form of iron instead of Fe^{2+} as iron salts. During the last decade there has been a widespread development of nanomaterials for both industrial and domestic use. When the dimensions of a piece of solid material become very small, its physical and chemical properties become very different from those of the same material in larger bulk form [24]. Nanoparticles as a subset of nanomaterial is currently defined by consensus as single particles with a diameter < 100 nm [24]. The small particle size increases the proportion of atoms located at the surface increasing the possibility for the atoms to adsorb, interact, and react with other atoms and molecules [25]. The particles have also the capacity to remain in suspension [25] and, hence, aqueous slurries containing nZVI can easily be pumped and injected where needed. The nZVI has been mostly used for groundwater remediation and treatment of specific pollutants [25]. Additionally, the treatment of industrial wastewater using nZVI in Fenton and other processes has also been reported recently [26, 27].

The main objective of this investigation was two-folded: (i) to verify the technical feasibility of treating wastewaters generated in the wood-floor industry sector using Fenton with nZVI compared to photo-Fenton with nZVI and (ii) to verify the effects of the selected variables (H_2O_2/COD ratio, H_2O_2/nZVI ratio and dosing mode) on the treatment efficiency.

MATERIALS AND METHODS

Wastewater Samples

The wastewater used in the current investigation was a mixture of different streams of real industrial wastewater generated during cleaning procedures in a wood-floor industry in Nybro, Sweden. The quality and quantity of these wastewaters vary in time, as they are manually and intermittently generated as a consequence of different manufacturing processes such as wood gluing, wood filling, cleaning of floors, blade sharpening, and others. These wastewaters are characterized by the presence of formaldehyde, nitrogen [1], metals [28] detergents, and, phenols [29]. At the factory, the wastewater mixture is kept in a full-scale on-site settling/sedimentation tank from where the samples for lab studies were taken. Wastewater samples were obtained at three different occasions and transported to the laboratory where they were stored at −20°C.

In order to obtain an average composition of the wastewater, a mixture of similar proportion of the three samples stored in the lab (1 : 1 : 1) was prepared. This mixture was then filtered with a Munktell OOR grade filter paper with a pore size >10 μm to homogenize the wastewater. The characteristics of the wastewater used in the experiments are shown in Table 1. Proxy indicators such as TOC and COD were considered suitable to monitor treatability studies of this wastewater, since it is well known that these variables as well as the ratio between them are appropriate to evaluate the efficiency of treatment options [30, 31].

Table 1: Wastewater characterization (n=3)

Parameter	Value	Mixture*
pH	2.3 ± 0.5	2.2
Conductivity (mS/cm)	6.3 ± 0.6	5.9
COD (mg/L)	5102 ± 513	4956

TOC (mg/L)	2,801 287	±	2730

*Wastewater obtained from a mixture of these three characterized samples.

Experimental Design

To have a better understanding on which independent variables play important roles on the treatment efficiency measured as percentage reduction of COD and TOC (dependent variables), it was applied a two-level full-factorial design with triplicates of the central points with the following selected independent variables or factors: (1) H_2O_2:COD ratio; (2) H_2O_2:nZVI ratio and; (3) the dosing mode. By dosing mode one means the procedure of adding equal aliquots of the oxidizing agent (H_2O_2) and the catalyst (nZVI) at different times throughout the experiment or adding it at once. The levels used for each factor in this investigation (Table 2) were selected according to the literature.

Table 2: Variable levels applied in the 2-level factorial design

Variables	Symbol	−1	0	+1
H_2O_2:COD	χ_1	2:1	3.5:1	5:1
H_2O_2:nZVI	χ_2	2:1	8.5:1	15:1
Dosing mode	χ_3	1	2	3

Notations: +1(high level); 0 (centre point); −1 (lower level).

The AOS (average oxidation state) value was calculated using the COD and TOC data obtained by the above-described experimental design. The AOS value can attain a value between +4, for CO_2, and −4, for CH_4, the most oxidized and the most reduced state of carbon (C). The AOS value is a rough parameter to estimate the degree of oxidation in a mixed wastewater and was calculated according to the following [31]:

$$AOS = \frac{4(TOC - COD)}{TOC},$$

(5)

where COD and TOC values are expressed in mol O_2/L and C/L, respectively.

The statistical software Minitab 16 was used to setup the full factorial design. For statistical analysis Minitab 16 and GraphPad Prism 5 were used.

Experimental Setup

Two variations of Fenton treatment were investigated: dark-Fenton (nZVI/H_2O_2) and photo-Fenton (nZVI/H_2O_2/UV), using, in both cases, commercial nZVI. The commercial nZVI powder consisted of Fe^0 surface stabilized nanoparticles coated with a thin inorganic surface layer, which makes it possible long-term storage as the manufacturer delivers it. The material had particle sizes ranging between 20–100 nm with an average of 50 nm according to the manufacturer (Nano Iron, s.r.o., Czech Republic). The photo-Fenton studies were carried out in a glass-jacket immersion-type UV-reactor (UV-Consulting Peschl Mainz, Germany) with a volume of 0.7 L (Figure 1). The distance between the UV-lamp and the liquid phase was 15 mm. The reactor was cooled down with distilled water to keep the liquid phase at room temperature of around 22°C. The UV-light was emitted at a 150 W in wavelengths ranging between 250 and 580 nm with the highest peaks at 310, 360, 400, 440, 550, and 580 nm.

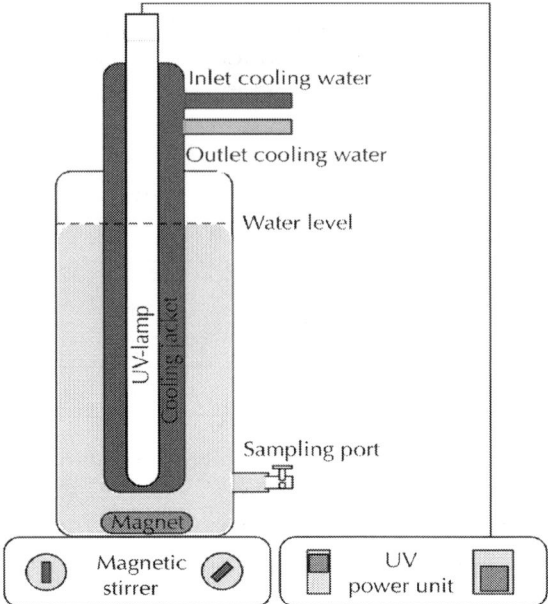

Inlet cooling water

Outlet cooling water

Water level

UV-lamp

Cooling jacket

Sampling port

Magnet

Magnetic stirrer

UV power unit

Figure 1: Schematic view of the UV-reactor used in the photo-Fenton experiments.

The dark-Fenton studies were conducted in 1 L glass beakers. The difference between standard glass beaker and the UV-reactor with the light in the off-mode was considered negligible according to the literature [32, 33]. Air purging was not considered in this study in order to keep treatment operational costs as low as possible. All glassware was carefully washed before each experiment.

Experimental Procedure

All runs were conducted with a volume of 0.5 L of wastewater that was agitated during 120 min by a magnetic stirrer at 400 rpm. The pH was adjusted at the beginning of each run to be kept between 2.95 and 3.05. Since this pH range has been reported to be the optimum for Fenton oxidation [13, 34], every 15 min throughout the 120 min of reaction time the pH was readjusted if needed. All pH adjustments were done with analytical grade sodium hydroxide (NaOH) and hydrochloric acid (HCl). The total amounts of nZVI and H_2O_2 were added either at the beginning of each run or in similar aliquots, at different time intervals (Table 3). The nZVI was added as slurry formed by $1:4$ nZVI powder and water, after being stirred in a high speed sheerer for 5 min, according to the manufacturer's instructions.

Table 3: Dosing modes (the way the H_2O_2 and nZVI were added during the runs)

Sample number*	Total amount of H_2O_2 (g/L)	Total amount of nZVI (g/L)	Number of equal doses	0 min		40 min		60 min		80 min		120 min
				H_2O_2 (g/L)	nZVI (g/L)	H_2O_2 (g/L)	nZVI (g/L)	H_2O_2 (g/L)	nZVI (g/L)	H_2O_2 (g/L)	nZVI (g/L)	
1.1; 2.1; 3.1; 4.1	24,8	1,7	1	24,8	1,7	—	—	—	—	—	—	—
1.2; 2.2; 3.2; 4.2	9,9	0,7	1	9,9	0,7	—	—	—	—	—	—	—
1.3; 2.3; 3.3; 4.3	17,3	2,0	2	8,7	1,0	—	—	8,7	1,0	—	—	—
1.4; 2.4; 3.4; 4.4	17,3	2,0	2	8,7	1,0	—	—	8,7	1,0	—	—	—
1.5; 2.5; 3.5; 4.5	9,9	0,7	3	3,3	0,2	3,3	0,2	—	—	3,3	0,2	—
1.6; 2.6; 3.6; 4.6	9,9	5,0	3	3,3	1,7	3,3	1,7	—	—	3,3	1,7	Full treatment time
1.7; 2.7; 3.7; 4.7	17,3	2,0	2	8,7	1,0	—	—	8,7	1,0	—	—	—
1.8; 2.8; 3.8; 4.8	24,8	1,7	3	8,3	0,6	8,3	0,6	—	—	8,3	0,6	—
1.9; 2.9;	24,8	12,4	3	8,3	4,1	8,3	4,1	—	—	8,3	4,1	—

3.9; 4.9□											
1.10 2.10 3.10 4.10□	24,8□	12, 4□	1□	24,8□	12 ,4□	—□	—□	—□	—□ —□	—□	□
1.11 2.11 3.11 4.11□	9,9□	5,0□	1□	9,9□	5, 0□	—□	—□	—□	—□ —□	—□	□

*Treatments from 1.1 to 1.11: dark-Fenton; from 2.1 to 2.11: Photo-Fenton. □

During the photo-Fenton treatment, the wastewater was exposed to UV-light during the treatment time of 120 min. After both Fenton and photo-Fenton treatments of 120 min, the pH of the wastewater was adjusted to 6.5 and agitated for 5 min at 400 rpm to quench the Fenton reaction [35]. In a sequence, 5 min after the agitation was stopped, the pH was raised to 8.5 to form iron precipitates [35]. The supernatant was then heated to 50°C and slowly shaken in a water bath for 30 min to expel any remaining H_2O_2. The water was then centrifuged at 614 g for 15 min. After this, the supernatant was separated and all samples were frozen before analysis.

Analytical Methods

COD and TOC in the wastewater samples were analysed spectrophotometrically using Hach Lange cuvette tests (Hach Lange, Dusseldorf) and measured with a Hach Lange DR 5000 spectrophotometer (Hach Lange, Dusseldorf). The pH and conductivity were measured with an HQ40d multiparameter meter.

RESULTS AND DISCUSSION

Comparison between Fenton and Photo-Fenton

nZVI Fenton Treatment

The treatments based on dark-Fenton (Table 4 and Figures 2(a) and 2(b)) showed large variation in the responses measured as the COD and TOC removal percentages. As observed in Table 4, the removal of COD and TOC varied from 34% to 77% and from 17% to 50%, respectively, depending on the run indicating that the ranges of $H_2O_2:COD$; $H_2O_2:nZVI$ and dosing modes combined in different ways have played an important role on the COD and TOC removal. There was a significant difference between COD and TOC reduction % within the dark-Fenton and photo-Fenton processes (paired t-test, $P<0.05$) as illustrated in Table 4.

Table 4: COD and TOC reductions (%) with dark-Fenton and photo-Fenton experiments

| Coded variables | | | | Dark-Fenton | | | Photo-Fenton | |
x_1	x_2	x_3	Run	% COD reduction	% TOC reduction	Run	% COD reduction	% TOC reduction
+1	+1	−1	1.1	42.6	21.4	2.1	78.4	59.4
−1	+1	−1	1.2	34.6	17.3	2.2	70.6	50.0
0	0	0	1.3	42.0	21.4	2.3	80.5	57.8
0	0	0	1.4	39.3	19.9	2.4	78.2	56.7
−1	+1	+1	1.5	35.6	19.0	2.5	77.6	56.0
−1	−1	+1	1.6	35.1	18.9	2.6	76.1	53.9
0	0	0	1.7	38.4	24.2	2.7	61.3	61.7
+1	+1	+1	1.8	43.5	23.0	2.8	60.3	61.7
+1	−1	+1	1.9	77.5	50.3	2.9	81.6	61.6
+1	−1	−1	1.10	49.2	34.9	2.10	78.5	58.0
−1	−1	−1	1.11	38.3	24.7	2.11	70.9	53.6

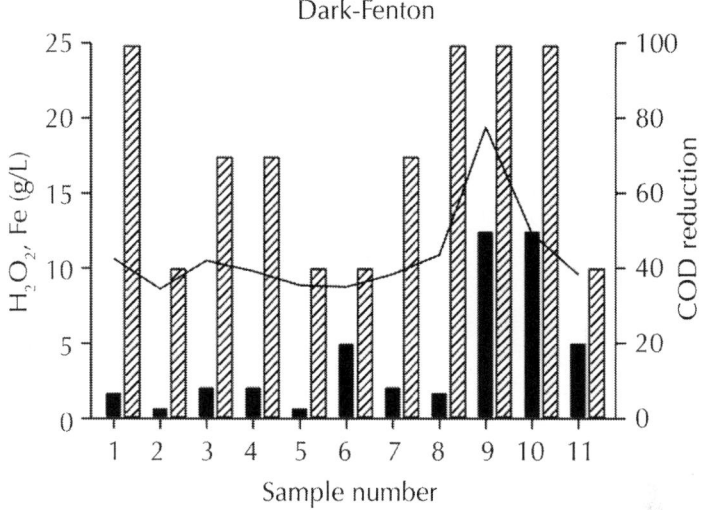

(a)

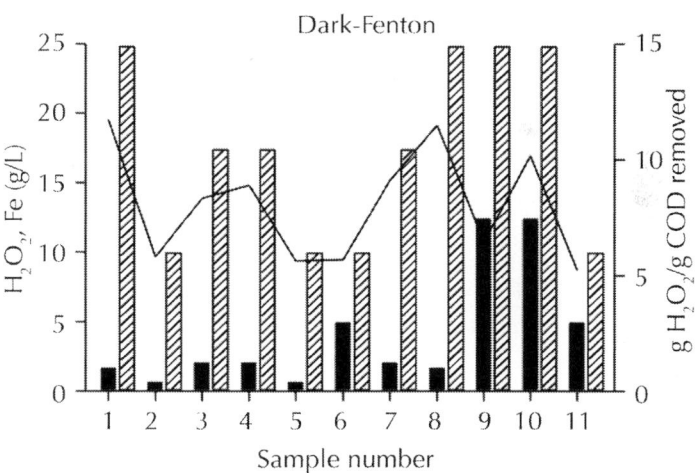

(b)

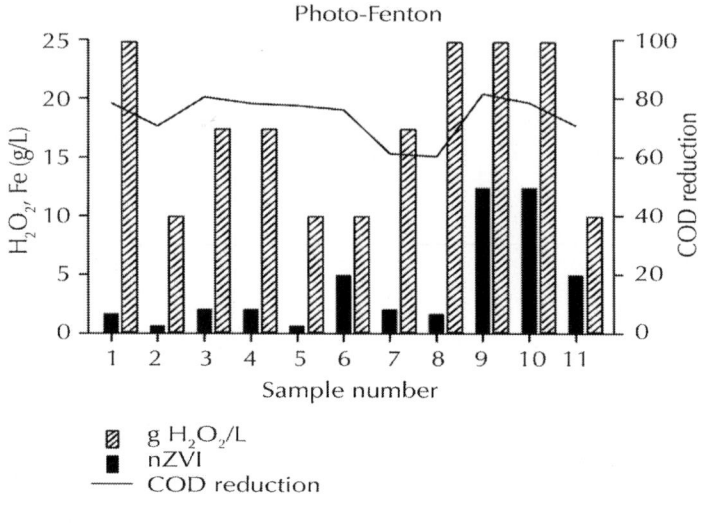

(c)

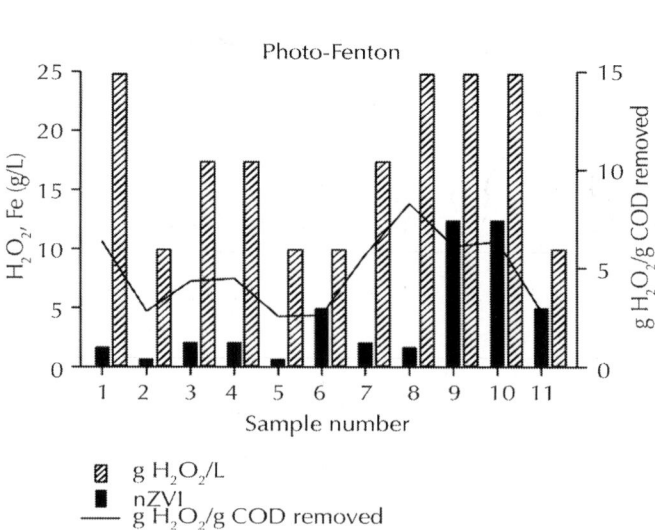

Figure 2: Results from dark-Fenton (a) and (b) and photo-Fenton (c) and (d) experiments using nZVI in 11 runs each. (a) and (c) Amounts of reagents (g/L) and COD removal (%); (b) and (d) amounts of reagents (g/L) and g H_2O_2 consumed per g COD removed.

The highest reductions for both COD and TOC were achieved with a setup using high H_2O_2 : COD ratios and low H_2O_2 : nZVI ratio (run 1.9 in Table 4 and 9 in Figure 2(a)), indicating that higher concentrations of the oxidizing agent (H_2O_2) and the catalyst were able to oxidize a higher amount of COD and TOC, despite the fact that high concentrations of nZVI will possibly scavenge hydroxyl radicals. The stoichiometric ratio for the reduction of COD by H_2O_2 is 2.125 which is calculated assuming the complete oxidation of COD [36]:(see [36]).

$$1 \text{ g COD} = 1 \text{ g O}_2 = 0.03125 \text{ mol O}_2$$

$$= 0.0625 \text{ mol H}_2O_2 = 2.125 \text{ g H}_2O_2, \quad (6)$$

The results have shown that the amount of H_2O_2 to achieve the best treatment efficiency was twice as much the amounts stoichiometrically required. However such occurrences can be due to several factors that need to be further investigated. An important aspect regarding the Fenton treatment is the reduction of Fe^{3+} to Fe^{2+} making crucial the presence of reaction intermediates able to reduce Fe^{3+} and regenerate the catalyst. However, there are reaction intermediates that instead of reducing the Fe^{3+} remove it from the Fe^{2+}/Fe^{3+} cycle, due to the generation of iron complexes, delaying and/or inhibiting the oxidation process [18]. Regardless the use of higher amounts of H_2O_2 in comparison to the stoichiometric need, the H_2O_2 : COD ratio of 5:1 was in a range similar to that reported as being effective in previous studies [14, 18]. On the other hand the H_2O_2 : nZVI ratio of 2:1 was lower than the values found in literature [12, 18, 37].

When analysing the amount of oxidizing agent H_2O_2 per unit of COD removed (g H_2O_2/g COD removed) used in different treatments in Figure 2(b), other treatments can be considered economically more feasible (runs 1.2, 1.5, 1.6, and 1.11 in Table 4; respectively 2, 5, 6, and 11 in Figure 2(b)). But run 1.9 was much more efficient in terms of COD removal and it was not considerably worse than the others above mentioned regarding g H_2O_2/g COD removed.

nZVI Photo-Fenton Treatment

The results regarding COD and TOC removals with photo-Fenton have shown that the addition of the UV energy has significantly increased the treatment efficiency compared to dark-Fenton (paired t-test, $P<0.05$)

(Figures 3(a) and 3(b)). Furthermore, it was observed that during the photo-Fenton process, the reductions of COD and TOC were more homogenous in comparison to those observed for the dark-Fenton treatment (Table 4; Figures 2(c) and 2(d)).

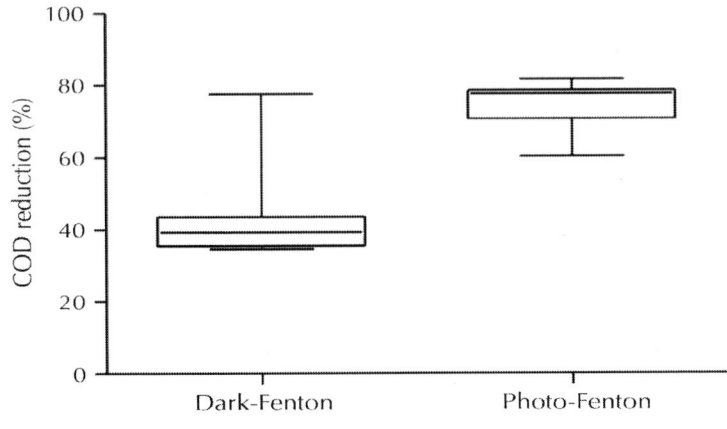

(a)

(b)

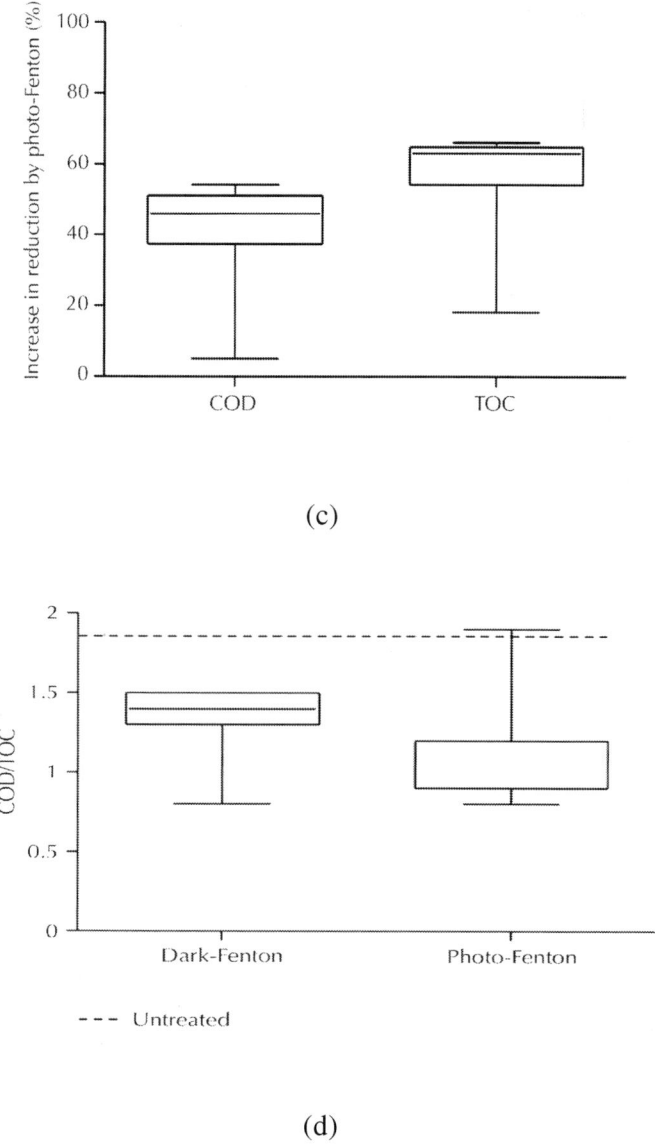

(c)

(d)

Figure 3: Boxplots built up with data from COD and TOC values obtained with dark-Fenton and photo-Fenton treatments. (a) COD reduction (in %) with both treatments; (b) TOC reduction (in %) with both treatments; (c) increase (in %) in the percentage of COD and TOC reduction by photo-Fenton compared to dark-Fenton; (d) changes in COD/TOC ratio due to different treatments.

The increase of TOC reduction was significantly higher than the increase of COD reduction by photo-Fenton (Figure 3(c)). A significant difference was observed between the reductions (in %) of COD and TOC with photo-Fenton treatment (paired t-test, $P<0.05$).

Run 2.9 in Table 4 (equivalent to 9 in Figure 2(c)) promoted the highest removal of COD and TOC in %. However, when taking into account the amount of oxidizing agent used per amount of COD removed (g H_2O_2/g COD removed), another treatment was economically more effective (run 2.5 in Table 4, equivalent to run 5 in Figure 2(d)). In Figure 2, both runs 5 and 6 performed satisfactorily and with low amounts of H_2O_2 promoted COD removals of around 80%, indicating the crucial role played by the UV energy in terms of treatment efficiency. The efficiency in using the UV in combination with the Fenton reaction is confirmed in Figures 2(c)and 2(d), which illustrate that despite using a H_2O_2 : COD ratio as high as 5 : 1 in run 2.9 (Table 4), only a small increase in COD removal in % was observed compared to runs 2.5 and 2.6 (Table 4) where a H_2O_2 : COD ratio of only 2 : 1 was used. COD and TOC reductions over 75% and 55%, respectively, were achieved in runs 2.5 and 2.6 (Table 4), respectively, which corresponded to less than half of the amount of H_2O_2applied in run 2.9 (COD and TOC reductions of 82% and 61%, resp.). Similar results were reported [38] where an increase of H_2O_2 : COD ratio above the range of 1–3 did not improve the COD reduction related to the presence of antibiotics in the water. Furthermore, the use of high concentrations of nZVI in runs 2.6 and 2.9 (H_2O_2 : nZVI of 2 : 1, Table 4) did not result in a increased efficiency, in comparison to run 2.5 (H_2O_2 : nZVI of 15 : 1). As discussed previously, one important event for the Fenton process is the reduction of Fe^{3+} to Fe^{2+} and such reduction is enhanced when using photo-Fenton [20]. This is likely one of the reasons for the much larger reductions in general achieved with photo-Fenton, being another reason for the fact that significantly lower concentrations of nZVI can be used. It is worth noting that in all these setups the reactants were added in three doses along the treatment, reducing the risk for scavenging between the reactants. Figure 2(d) shows that when 94% less nZVI was used, only 6% less COD and TOC reductions were observed in the presence of UV. Such a large reduction of iron reduces the costs with this catalyst and the amount of spent iron that needs to be handled (see the discussion Cost Effectiveness in this paper).

Oxidation and Mineralisation

The results have shown that the COD and TOC values after both dark- and photo-Fenton treatments were considerably different as indicated by the COD/TOC ratios (Figure 3(d)). There was no significant difference in COD/TOC ratios when comparing photo-Fenton and dark-Fenton; however, COD/TOC ratios decreased after treatment. As illustrated in Figure 4, whereas negative correlations were found between COD and TOC reduction (in %) and COD/TOC ratio (R^2=0.86 and 0.62, resp.) for the dark-Fenton experiments, only COD removal (in %) was correlated with COD/TOC ratio (R^2=0.92) in the photo-Fenton investigation. The correlation between TOC removal (in %) and COD/TOC for photo-Fenton had coefficient of determination as low as R^2=0.18. These results suggest that the remaining organic compounds measured as TOC were very recalcitrant and difficult to degrade as indicated by very low COD/TOC ratios after these treatments, and even with higher reductions of COD, the TOC value was not lowered. Previous studies have reported that a COD/TOC ratio below 1.3 indicates that that residual organic carbon was mostly related to refractory organic compounds [16].

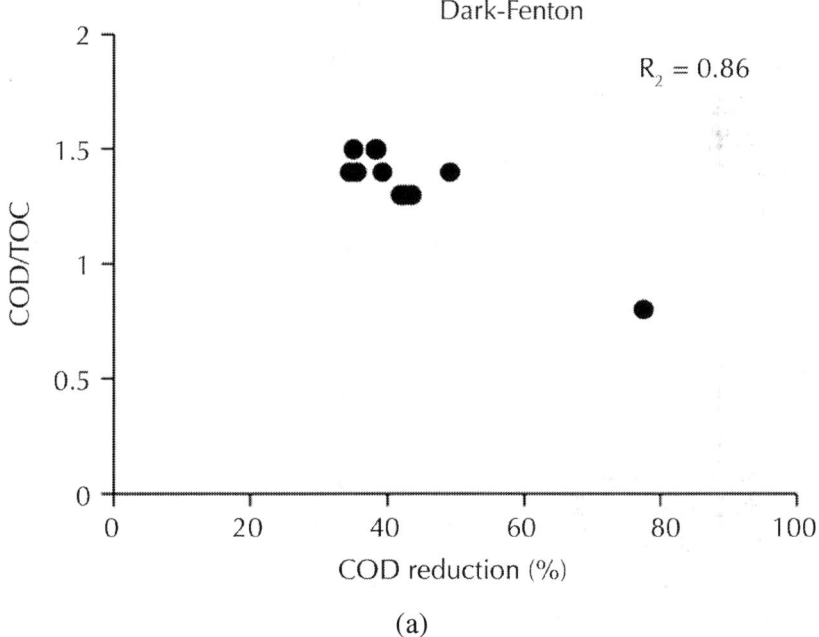

(a)

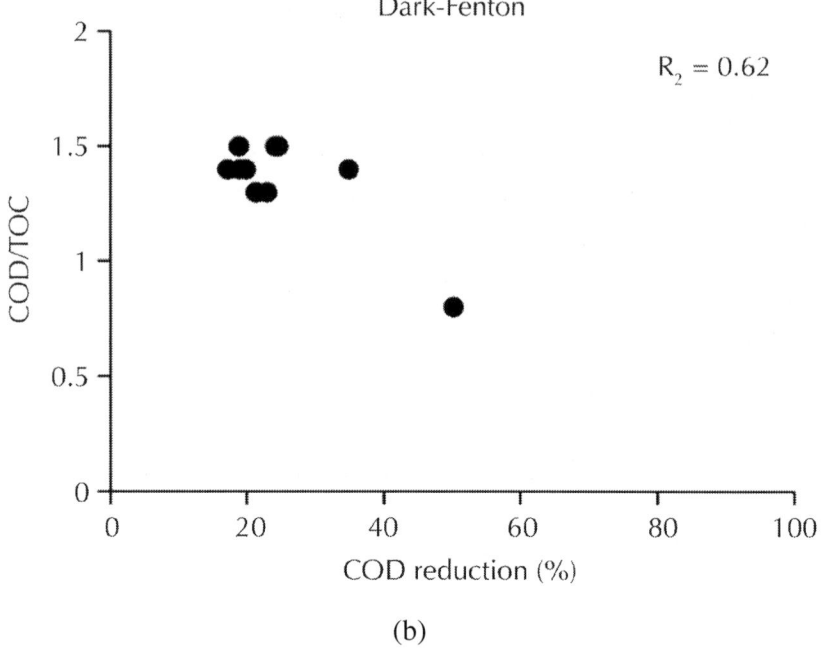

(b)

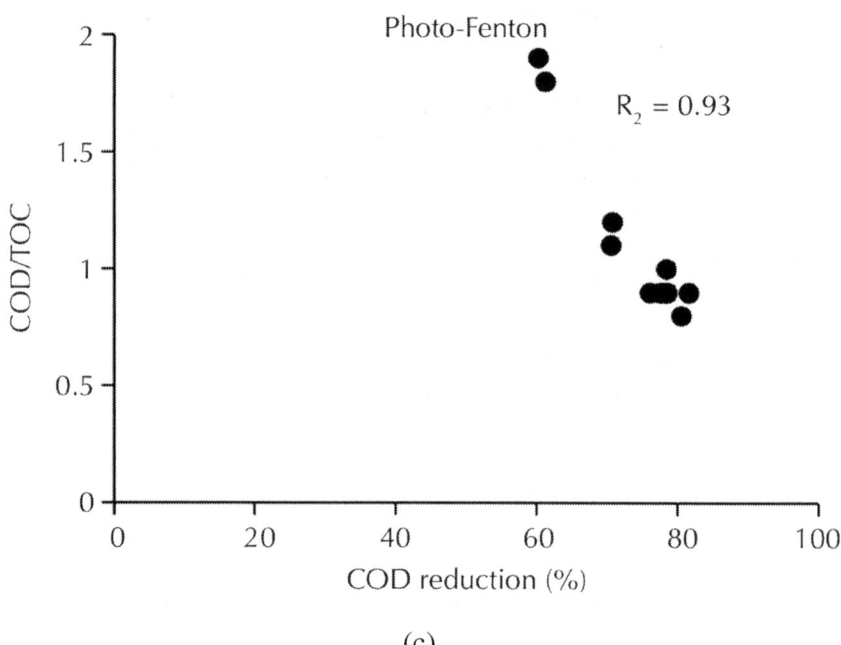

(c)

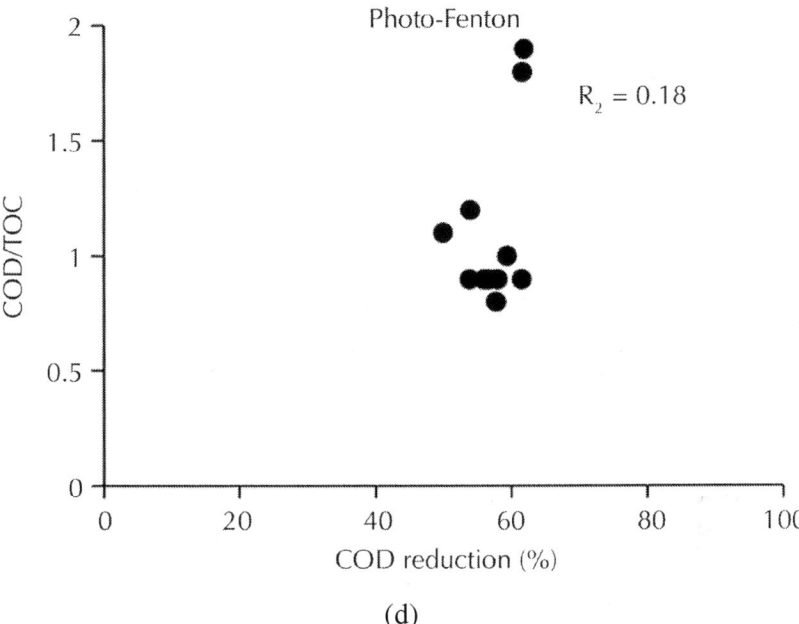

(d)

Figure 4: Correlation between COD and TOC reductions (in %) and COD/TOC ratios obtained with different dark- and photo-Fenton treatment setups.

Unlike the COD value that is related to organically-bound and inorganic constituents [39], TOC is independent of the oxidation state of the organic matter and only measures organic carbon converted to CO_2 [40], suggesting that changes observed in the COD/TOC ratio can be related to the degree of changes in the structure of the organic compounds after oxidation. Figure 5 shows the AOS values before and after the treatment with dark-Fenton and photo-Fenton. It was found that by the end of the photo-Fenton treatment, the oxidation states were higher in comparison to those after the dark-Fenton treatment, confirming the stronger oxidation (higher reduction efficiency for COD and TOC) with photo-Fenton and suggesting the yield of different end products. Regarding the oxidation states in the dark-Fenton, the AOS value in the run 1.9 was considerably higher in comparison to the raw wastewater and to other dark-Fenton treatments reaching AOS value approximately as high as those obtained after the photo-Fenton process. According to Figure 5, with the exception of runs 2.7 and 2.8, AOS values after the photo-Fenton treatments were approximately

+3 in all remaining runs suggesting that the oxidised chemical nature of most of the photo-Fenton treatments reached a stable AOS value and even though the treatment was to be continued, intermediates with higher oxidation state would not be formed. At the moment that AOS stabilizes, the chemical treatment is only mineralizing organic contaminants, but with no partial oxidation [41]. This together with the fact that the difference between the COD and TOC after treatment was considerably reduced by the photo-Fenton treatment (paired t-test, P<0.05) indicates that a more complete reduction of the organic material was achieved when applying the photo-Fenton process. An increase in AOS during the treatment as observed in the current study, particularly for the photo-Fenton treatments, has been related to an increase in biodegradability [41] and reduction of toxicity [38].

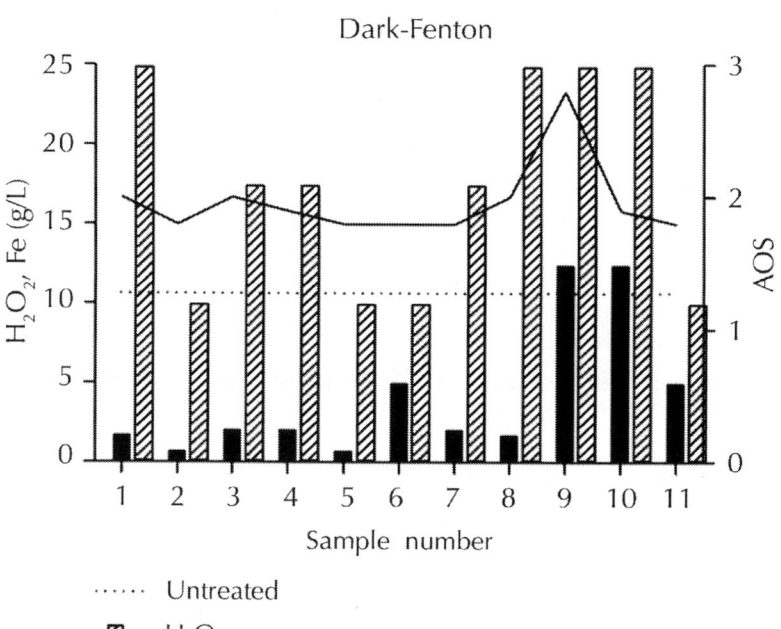

(a)

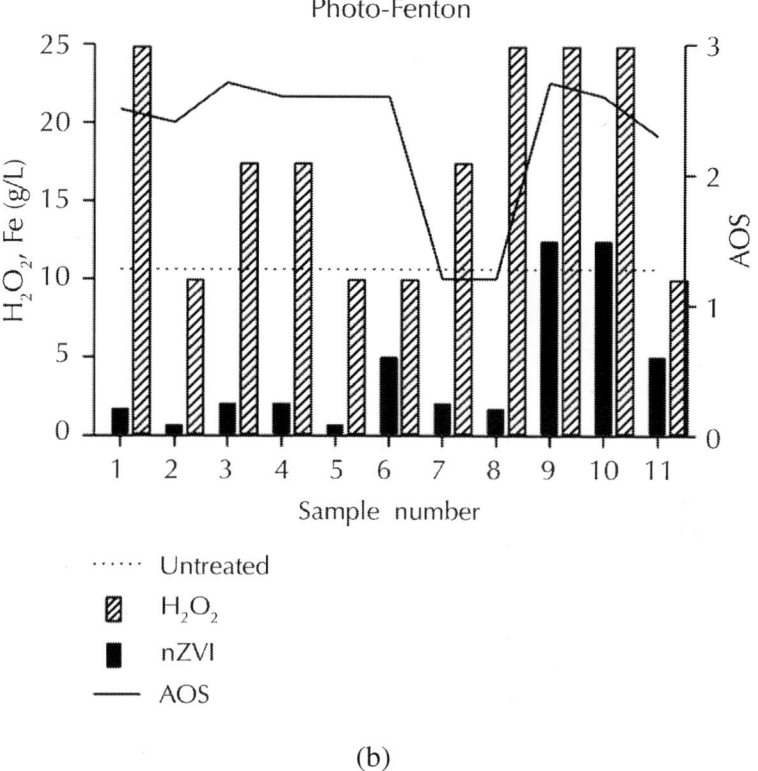

(b)

Figure 5: Amounts of reagents (g/L) and the AOS values obtained after each run/treatment.

This was confirmed by the fact that the reduction (in %) obtained with photo-Fenton is greater for TOC than for COD (paired t-test, $P < 0.05$), (Figure 3(c)). UV irradiation might have been responsible, for instance, for breaking down large aromatic molecules into aliphatic carbon chains, besides its effect on recycling Fe^{3+} back to Fe^{2+} [20].

Cost Effectiveness

Even though economic feasibility was not the focus of this investigation, one can mention that the main costs involved in the studied processes are related to the nZVI, the oxidizing agent H_2O_2, and the energy consumption of the UV-lamp. Photo-Fenton is reasonably more expensive than dark-Fenton treatment due to the use of a UV-lamp.

However, the results have shown that when considering the overall treatment efficiency of both Fenton and photo-Fenton, the additional costs related to the UV-lamp could be compensated by the less amounts of nZVI to achieve high COD and TOC reduction %. Furthermore, the use of UV energy has brought advantages since it widened the range of reactant concentrations that could be used and still achieve very satisfactory treatment performances, making reasonable to state that is the less the amount of reactants, the more economically feasible the treatment becomes. Such advantage becomes very important in a full scale plant, where the concentration of pollutants in the raw wastewater can vary, which would require the use of different concentrations of reactants. A clear example is the run 2.5 (Table 4), which had the most satisfactory performance observed with the photo-Fenton treatment. In run 2.5 with the lowest reactant concentrations, COD and TOC reductions of 78% and 56%, respectively, were achieved, making this the most economically feasible option in this study.

Effect of Independent Variables

The effects of the independent variables (H_2O_2 : COD; H_2O_2 : nZVI and dosing mode) and the interaction among them on the COD and TOC removal (in %) were obtained based on the data from the full-factorial design as shown in Table 5. Whereas all factors and respective two-way and three-way interactions had significant effects on the COD and TOC removal efficiency with dark-Fenton process ($P<0.05$), no significance was found in the case of photo-Fenton (Table 5). Reasonably, H_2O_2 : COD and H_2O_2 : nZVI ratios were the factors that played the most important roles considering dark-Fenton treatment. Reasonably, negative values were obtained for the effects caused by the H_2O_2 : nZVI ratios on COD and TOC reduction in % with dark-Fenton, since the lower the ratio is, the higher the treatment efficiency will be as previously discussed. The results have shown that by reducing the H_2O_2 : nZVI ratio from 15 : 1 down to 2 : 1, the reductions (in %) of COD and TOC could be increased by 11% and 12%, respectively. The significant effect of the dosing mode on both COD and TOC reductions suggests that when large amounts of reactants are added at once, the treatment performance was not increased due to scavenging [37, 42]. On the other hand, by splitting the doses of the reactants into several steps, scavenging was avoided, an effect that has been described only very few previous studies [42, 43]. The results have shown that the studied factors and levels applied did not play any significant role in the case of photo-Fenton experiments. These results suggest that the long UV-exposure time has eliminated large differences in the results or by combining chemical oxidation and UV irradiation other factors might be involved.

Table 5: Results of statistical analyses of independent variables

Term	Dark-Fenton				Photo-Fenton			
	COD reduction		TOC reduction		COD reduction		TOC reduction	
	Effect	P	Effect	P	Effect	P	**Effect**	P
Constant		0.00		0.00		0.00		0
H_2O_2:COD Ratio	17.3	0.01	12.4	0.02	0.9	0.91	6.8	0.07
H_2O_2:nZVI Ratio	−11.0	0.02	−12.0	0.02	−5.0	0.57	0.0	0.99
Doses	6.8	0.04	3.3	0.12	−0.7	0.94	3.0	0.24
H_2O_2:COD Ratio * H_2O_2: nZVI Ratio	−9.3	0.02	−8.4	0.03	−5.7	0.52	0.9	0.68
H_2O_2:COD Ratio * Dose	7.8	0.03	5.2	0.08	−6.8	0.46	0.0	0.99
H_2O_2:nZVI Ratio * Dose	−5.8	0.05	−1.6	0.40	−4.9	0.58	1.2	0.57
H_2O_2:COD Ratio * H_2O_2:nZVI Ratio*Dose	−8.0	0.02	−5.3	0.07	−5.7	0.52	−1.8	0.43
Ct Pt		0.07		0.01		0.91		0.39

CONCLUSIONS

The following conclusions can be withdrawn from this investigation:

* COD and TOC can effectively be reduced by at least 80% and 60%, respectively, from recalcitrant industrial wastewater from the wood industry using photo-Fenton

- the most effective treatment setup for dark-Fenton was achieved with H_2O_2 : COD ratio of 5 : 1, H_2O_2 : nZVI ratio of 2 : 1 and with a dose mode that supplies the reactants in 3 equal aliquots added at equal time intervals

- the most effective treatment setup for photo-Fenton was obtained with H_2O_2 : COD ratio of 5 : 1, H_2O_2 : nZVI ratio of 2 : 1, supplying the reactants in 3 equal aliquots added at equal time intervals; however, treatments close to the stoichiometry value (2.125) preformed almost as good

- there was a significant increase in the mineralization when combining UV with Fenton (photo-Fenton).

ACKNOWLEDGMENTS

The financial support to the research project from the Swedish Knowledge Foundation (KK-Stiftelsen), the European Regional Development Fund, and the industry AB Gustaf Kähr is acknowledged.

REFERENCES

1.   F. Kaczala, M. Marques, and W. Hogland, "Biotreatability of wastewater generated during machinery washing in a wood-based industry: COD, formaldehyde and nitrogen removal," Bioresource Technology, vol. 101, no. 23, pp. 8975–8983, 2010.

2.   S. Laohaprapanon, M. Marques, and W. Hogland, "Removal of organic pollutants from wastewater using wood fly ash as a low-cost sorbent," Clean, vol. 38, no. 11, pp. 1055–1061, 2010. · ·

3.   M. Krzemieniewski, M. Debowski, A. Dobrzynska, and M. Zielinski, "Chemical oxygen demand reduction of various wastewater types using magnetic field-assisted fenton reaction," Water Environment Research, vol. 76, no. 4, pp. 301–309, 2004.
·

4.   A. Kyriacou, K. E. Lasaridi, M. Kotsou, C. Balis, and G. Pilidis, "Combined bioremediation and advanced oxidation of green table olive processing wastewater," Process Biochemistry, vol. 40, no. 3-4, pp. 1401–1408, 2005. · ·

5. D. Suryaman, K. Hasegawa, and S. Kagaya, "Combined biological and photocatalytic treatment for the mineralization of phenol in water,"Chemosphere, vol. 65, no. 11, pp. 2502–2506, 2006. · ·

6. I. Oller, S. Malato, J. A. Sánchez-Pérez, M. I. Maldonado, and R. Gassó, "Detoxification of wastewater containing five common pesticides by solar AOPs-biological coupled system," Catalysis Today, vol. 129, no. 1-2, pp. 69–78, 2007. ·

7. W. K. Lafi, B. Shannak, M. Al-Shannag, Z. Al-Anber, and M. Al-Hasan, "Treatment of olive mill wastewater by combined advanced oxidation and biodegradation," Separation and Purification Technology, vol. 70, no. 2, pp. 141–146, 2009. · ·

8. O. Tunay, et al., Chemical Oxidation Applications for Industrial Wastewaters, IWA, 2010.

9. F. Torrades, S. Saiz, and J. A. García-Hortal, "Using central composite experimental design to optimize the degradation of black liquor by Fenton reagent," Desalination, vol. 268, no. 1–3, pp. 97–102, 2011. · ·

10. K. Choi and W. Lee, "Enhanced degradation of trichloroethylene in nano-scale zero-valent iron Fenton system with Cu(II)," Journal of Hazardous Materials, vol. 211-212, pp. 146–153, 2012.

11. M. I. Badawy, M. Y. Ghaly, and T. A. Gad-Allah, "Advanced oxidation processes for the removal of organophosphorus pesticides from wastewater,"Desalination, vol. 194, no. 1–3, pp. 166–175, 2006. · ·

12. I. Arslan-Alaton Idil, A. B. Yalabik, and T. Olmez-Hanci, "Development of experimental design models to predict Photo-Fenton oxidation of a commercially important naphthalene sulfonate and its organic carbon content," Chemical Engineering Journal, vol. 165, no. 2, pp. 597–606, 2010. · ·

13. K. V. Padoley, S. N. Mudliar, S. K. Banerjee, S. C. Deshmukh, and R. A. Pandey, "Fenton oxidation: a pretreatment option for improved biological treatment of pyridine and 3-cyanopyridine plant wastewater," Chemical Engineering Journal, vol. 166, no. 1, pp. 1–9, 2011. · ·

14. D. Hermosilla, N. Merayo, R. Ordóñez, and A. Blanco, "Optimization of conventional Fenton and ultraviolet-assisted oxidation processes for the treatment of reverse osmosis retentate

from a paper mill," Waste Management, vol. 32, no. 6, pp. 1236–1243, 2012.

15. P. Bautista, A. F. Mohedano, M. A. Gilarranz, J. A. Casas, and J. J. Rodriguez, "Application of Fenton oxidation to cosmetic wastewaters treatment," Journal of Hazardous Materials, vol. 143, no. 1-2, pp. 128–134, 2007. · ·

16. A. M. F. M. Guedes, L. M. P. Madeira, R. A. R. Boaventura, and C. A. V. Costa, "Fenton oxidation of cork cooking wastewater—overall kinetic analysis," Water Research, vol. 37, no. 13, pp. 3061–3069, 2003. · ·

17. M. Umar, H. A. Aziz, and M. S. Yusoff, "Trends in the use of Fenton, electro-Fenton and photo-Fenton for the treatment of landfill leachate," Waste Management, vol. 30, no. 11, pp. 2113–2121, 2010. · ·

18. P. Bautista, A. F. Mohedano, J. A. Casas, J. A. Zazo, and J. J. Rodriguez, "An overview of the application of Fenton oxidation to industrial wastewaters treatment," Journal of Chemical Technology and Biotechnology, vol. 83, no. 10, pp. 1323–1338, 2008.

19. W. Z. Tang and R. Z. Chen, "Decolorization kinetics and mechanisms of commercial dyes by H_2O_2/iron powder system," Chemosphere, vol. 32, no. 5, pp. 947–958, 1996. · ·

20. M. Tamimi, S. Qourzal, N. Barka, A. Assabbane, and Y. Ait-Ichou, "Methomyl degradation in aqueous solutions by Fenton›s reagent and the photo-Fenton system," Separation and Purification Technology, vol. 61, no. 1, pp. 103–108, 2008. · ·

21. I. Arslan and I. A. Balcioglu, "Advanced oxidation of raw and biotreated textile industry wastewater with O3, H_2O_2/UV-C and their sequential application," Journal of Chemical Technology & Biotechnology, vol. 76, no. 1, pp. 53–60, 2001.

22. B.-H. Moon, Y.-B. Park, and K.-H. Park, "Fenton oxidation of Orange II by pre-reduction using nanoscale zero-valent iron," Desalination, vol. 268, no. 1–3, pp. 249–252, 2011. · ·

23. M. Barreto-Rodrigues, F. T. Silva, and T. C. B. Paiva, "Combined zero-valent iron and fenton processes for the treatment of Brazilian TNT industry wastewater," Journal of Hazardous Materials, vol. 165, no. 1–3, pp. 1224–1228, 2009. · ·

24. P. J. A. Borm, D. Robbins, S. Haubold et al., "The potential risks of nanomaterials: a review carried out for ECETOC," Particle and Fibre Toxicology, vol. 3, no. 1, article 11, 2006. · ·

25. R. A. Crane and T. B. Scott, "Nanoscale zero-valent iron: future prospects for an emerging water treatment technology," Journal of Hazardous Materials, vol. 211-212, pp. 112–125, 2012.

26. S. Jagadevan, M. Jayamurthy, P. Dobson, and I. P. Thompson, "A novel hybrid nano zerovalent iron initiated oxidation— biological degradation approach for remediation of recalcitrant waste metalworking fluids," Water Research, vol. 46, no. 7, pp. 2395–2404, 2012.

27. Q. J. Rasheed, K. Pandian, and K. Muthukumar, "Treatment of petroleum refinery wastewater by ultrasound-dispersed nanoscale zero-valent iron particles," Ultrasonics Sonochemistry, vol. 18, no. 5, pp. 1138–1142, 2011. · ·

28. F. Kaczala, M. Marques, and W. Hogland, "Lead and vanadium removal from a real industrial wastewater by gravitational settling/ sedimentation and sorption onto Pinus sylvestris sawdust," Bioresource Technology, vol. 100, no. 1, pp. 235–243, 2009. ·

29. S. Laohaprapanon, F. Kaczalaa, P. S. Salomonbc, M. Marquesad, and W. Hoglanda, "Wastewater generated during cleaning/ washing procedures in a wood-floor industry: toxicity on the microalgae Desmodesmus subspicatus,"Environmental Technology. In press. ·

30. Z. Shiyun, Z. Xuesong, and L. Daotang, "Ozonation of naphthalene sulfonic acids in aqueous solutions. Part I: elimination of COD, TOC and increase of their biodegradability," Water Research, vol. 36, no. 5, pp. 1237–1243, 2002. · ·

31. V. Sarria, S. Kenfack, O. Guillod, and C. Pulgarin, "An innovative coupled solar-biological system at field pilot scale for the treatment of biorecalcitrant pollutants," Journal of Photochemistry and Photobiology A, vol. 159, no. 1, pp. 89–99, 2003. ·

32. F. J. Benitez, F. J. Real, J. L. Acero, C. Garcia, and E. M. Llanos, "Kinetics of phenylurea herbicides oxidation by Fenton and photo-Fenton processes,"Journal of Chemical Technology and Biotechnology, vol. 82, no. 1, pp. 65–73, 2007. · ·

33. H. Kusic, N. Koprivanac, and L. Srsan, "Azo dye degradation using Fenton type processes assisted by UV irradiation: a kinetic study," Journal of Photochemistry and Photobiology A, vol. 181, no. 2-3, pp. 195–202, 2006. · ·

34. C.-C. Su, M. Pukdee-Asab, C. Ratanatamskulb, and M.-C. Lu, "Effect of operating parameters on decolorization and COD removal of three reactive dyes by Fenton›s reagent using fluidized-bed reactor," Desalination, vol. 278, no. 1–3, pp. 211–218, 2011.

35. X. Zhu, J. Tian, R. Liu, and L. Chen, "Optimization of Fenton and electro-Fenton oxidation of biologically treated coking wastewater using response surface methodology," Separation and Purification Technology, vol. 81, no. 3, pp. 444–450, 2011.

36. M. S. Lucas and J. A. Peres, "Removal of COD from olive mill wastewater by Fenton›s reagent: kinetic study," Journal of Hazardous Materials, vol. 168, no. 2-3, pp. 1253–1259, 2009. · ·

37. C. T. Benatti, C. R. G. Tavares, and T. A. Guedes, "Optimization of Fenton›s oxidation of chemical laboratory wastewaters using the response surface methodology," Journal of Environmental Management, vol. 80, no. 1, pp. 66–74, 2006. · ·

38. E. Elmolla and M. Chaudhuri, "Optimization of Fenton process for treatment of amoxicillin, ampicillin and cloxacillin antibiotics in aqueous solution,"Journal of Hazardous Materials, vol. 170, no. 2-3, pp. 666–672, 2009. · ·

39. H. Zhang, J. C. Heung, and C. P. Huang, "Optimization of Fenton process for the treatment of landfill leachate," Journal of Hazardous Materials, vol. 125, no. 1–3, pp. 166–174, 2005. · ·

40. D. R. Medley and E. L. Stover, "Effects of ozone on the biodegradability of biorefractory pollutants," Journal of the Water Pollution Control Federation, vol. 55, no. 5, pp. 489–494, 1983. ·

41. C. Sirtori, A. Zapata, I. Oller, W. Gernjak, A. Agüera, and S. Malato, "Decontamination industrial pharmaceutical wastewater by combining solar photo-Fenton and biological treatment," Water Research, vol. 43, no. 3, pp. 661–668, 2009. · ·

42. Y. Deng and J. D. Englehardt, "Treatment of landfill leachate by the Fenton process," Water Research, vol. 40, no. 20, pp. 3683–3694, 2006. · ·

43. R. C. Martins, A. F. Rossi, and R. M. Quinta-Ferreira, "Fenton›s oxidation process for phenolic wastewater remediation and biodegradability enhancement," Journal of Hazardous Materials, vol. 180, no. 1–3, pp. 716–721, 2010. · ·

Zeolite Synthesis under Insertion of Silica Rich Filtration Residues from Industrial Wastewater Reconditioning

Andrea Hartmann[1], V. Petrov[1], J.-C. Buhl[1],
K. Rübner[2], M. Lindemann[2], C. Prinz[3],
and A. Zimathies[3]

[1]Institut für Mineralogie, Universität Hannover, Hannover, Germany

[2]Fachgruppe VII.1 Baustoffe, Bundesanstalt für Materialforschung und -prüfung (BAM), Berlin, Germany

[3]Fachbereich 1.3 Strukturanalytik Richard-Willstätter, Bundesanstalt für Materialforschung und -prüfung (BAM), Berlin, Germany

ABSTRACT

Zeolite synthesis was studied using two silica rich filtration residues (FR 1 and FR 2) as Si-source and sodium aluminate in a direct synthesis at 60°C at strong alkaline conditions (8 M - 16 M NaOH). In addition to these one-pot syntheses, a two-step process was investigated. Here, an alkaline digestion of FR at 60°C was followed by gel precipitation with sodium aluminate and gel crystallization under usual conditions of 80°C - 90°C. The results show that the substitution of chemical reagent sodium silicate by a waste material like FR as Si-source is possible but requires fine tuning of the reaction conditions as zeolite crystallization is a process under kinetic control. The solubility behaviour and impurities of the inserted filtration residues strongly influenced the course of reaction. Thus zeolites like hydrosodalite or intermediate zeolite between cancrinite and sodalite, or zeolite NaA or Z-21 in cocrystallization with hydrosodalite could be observed in the one pot syntheses already in a short time interval between 1 - 4 h depending on the alkalinity. The two step process yield to zeolites NaA and NaX in very good quality. The reaction process of FR in both reaction methods was characterized by chemical analyses, X-ray powder diffraction, Fourier transform infrared spectroscopy as well as scanning electron microscopy. Surface area and water content of selected products were further characterized by the BET-method and by thermogravimetry. Summing up the results, we can show that zeolite formation from filtration residues is possible by several reaction procedures as model cases for a re-use of industrial waste materials. Beside the importance for environmental protection, the reactions are of interest for zeolite chemistry as the re-use of FR is possible under economically conditions of low energy consumption at 60°C and short reaction periods.

INTRODUCTION

Zeolites are very important microporous materials, widely used in industry as sorbents, ion exchangers and catalysts. Most of the zeolites have aluminosilicate framework structures with different pore openings like the famous zeolites LTA and X. The outstanding zeolite LTA ("Linde Type A" structure) exhibits 4.1 Å sized pores in its sodium form and is mainly inserted as water softener in detergents. The "Faujasite" structure type zeolite Na-X successful used in catalysis has wider pores of 7.4 Å [1] - [3]. Usually zeolites were synthesized from alkaline aluminosilicate gels obtained by mixing of sodium silicate and sodium aluminate solutions. The gels were crystallized under mild hydrothermal conditions around 80°C for several hours [2] [3].

The chemical composition of gels for syntheses of important aluminosilicate zeolites is settled in the systems Na_2O-Al_2O_3-SiO_2-H_2O; K_2O-Al_2O_3-SiO_2-H_2O and CaO-Al_2O_3-SiO_2-H_2O. Beside pure chemicals even many industrial waste materials are suitable educts for zeolite formation. Models of use of adequate industrial wastes in zeolite chemistry would be of interest for waste management and environmental protection as well as a cheaper production under the premise of insertion of waste materials.

Materials like fly ash or slag with high concentrations of silica, aluminum oxide as well as an alkali (Na, K) and/or alkaline earth (Ca) content were already inserted in zeolite crystallization [4] -[9] . Synthesis of zeolites from fly ash of composition 50% by mass SiO_2 und 31% by mass Al_2O_3 and temperatures up to 250°C in 0.1 - 4.0 M NaOH, 0.1 - 4.0 M KOH and 0.1 - 1.0 M NaCl, KCl und $CaCl_2$ was investigated in [5] . Crystallization of analcite, Na-chabasite, K-chabasite, Faujasite and Phillipsit was observed under these conditions [5].

Zeolite LTA was successful synthesized from alumina rich slag in [10] by a more complex process. Starting with a melting step, followed by hydrothermal treatment to obtain sodium aluminum silicate as intermediate product finally crystallization was performed in NaOH solution for 1 - 9 h at 90°C [10].

As a contribution to investigate the applicability of further industrial waste materials beside slags or fly ash the present paper gives an experimental study on the insertion of filtration residues (FR) from waste water conditioning of production of silica, silane and zeolites. Those raw materials always exhibit high concentrations of SiO_2 thus being proper educts for a re-use in zeolite production. The large amounts of those residues incurred in industries, mentioned above, were mostly only deposited up to day and our aim is to show possible ways of its conversion into zeolites as a model case for recovery of its valuable ingredients. The insertion of waste material for zeolite production is of interest for environmental protection as well as for zeolite chemistry to develop economically procedures by re-use of its own production residues. A substitution of chemical reagent sodium silicate as Si-source in zeolite industry by a waste material like FR is not as simple as one would expect because zeolite crystallization is a process under kinetic control. Phases beyond the thermodynamic equilibrium will result and the solubility behaviour of the educts or an insertion of structure directing agents is simply influencing nucleation and growth and thus the complete phase formation process [11] - [13]. Hence the synthesis results are more or less depending on the condition of the educts especially its composition, solubility behaviour and content of impurities.

Thus, the following principles of reaction were investigated in the present paper under substitution of chemical grade Na_2SiO_3 by filtration residues of different composition (FR 1 and FR 2). In a first series the possibility of direct formation of zeolites from FR in a so called "one pot synthesis" is studied at low temperature under strong alkaline conditions, first described in [14] - [16] for pure chemicals gels. The temperature of only 60°C and the simple "one pot" procedure would discharge this method in a very economically process under low energy consumption. As this synthesis requires superalkaline conditions, reactions under total $Na_2O:H_2O$ ratios equivalent to 8 M, 12 M and 16 M NaOH in the complete reaction batch are investigated The conversion of the educts was studied in a short time interval between 1 - 4 h in dependence of the alkalinity. The results of this early period were compared with synthesis products obtained at a prolonged reaction time of 12 h. Besides time dependent investigation of phase formation, an influence on crystal size and morphology was also found.

In a second series, a two-step synthesis route is developed, starting with alkaline digestion of FR again at low temperature of only 60°C under 3 h digestion time. The digestion step was followed by common gel formation and crystallization under addition of $NaAlO_2$ at temperatures of 353 K or 363 K and times up to 16 h as reveled in crystallization of industrial zeolite LTA and X [1] -[3] .

The reaction process was characterized by chemical analyses using inductively coupled plasma optical emission spectrometry (ICP-OES) and by X-ray powder diffraction (XRD), Fourier transform infrared spectroscopy (FTIR) as well as scanning electron microscopy (SEM). Selected products were further characterized by the BET-method and by thermogravimetry.

The aims of this present experimental study are to investigate the mechanism and kinetics of zeolite formation as model case of a re-use procedure of filtration residue waste material (FR) by low energy consumption process at 60°C and short times. The study of conversion of FR in the early period of reaction and a control of phase stabilities at elevated times will further form the basis to clarify the question on occurrence of parallel and follow up reactions in dependence of alkalinity and time at low reaction temperatures.

In addition, a characterization of the zeolite products from FR by BET-surface area and thermogravimetry in comparison with literature data will show if qualitative sufficient products can be expected from a re-use of the FR waste materials.

EXPERIMENTAL

The Starting Raw Materials FR 1 and FR 2

Chemical analysis of FR 1 and FR 2 was performed by inductively coupled plasma optical emission spectroscopy (ICP-OES) using an iCAP 6000 by Thermo Scientific. The powders were totally decomposed in an acid mixture (nitric acid, phosphoric acid, fluoroboric acid) by means of microwave digestion. The results of analyses of the main components are summarized in Table 1.

The amorphous character of the filtration residues was demonstrated by X-ray powder diffraction (XRD) of FR 1 and FR 2 (equipment and conditions of measurements see analytical methods below). The powder patterns of the raw materials are given in Figure 1. Few parts of crystalline quartz can be seen in the powder pattern of FR 2 according the slight but intense main reflections of SiO_2 at 20.9°, 26.6° and 50.1° 2 θ in accordance with PDF46-1045 [17], whereas FR 1 is totally amorphous.

SEM investigations reveal the very fine grain size of both waste materials (see Figure 2, SEM results below). Nanoparticles around 100 - 200 nm, agglomerated to bigger aggregates of 2 μm and more were found for FR 1 and FR 2.

The Experimental Routes: Direct One Pot Synthesis and Two Step Synthesis

Direct syntheses with the filtration residues FR 1 and FR 2 were performed under magnetic stirring in 100 ml beakers at 60°C. Sodium aluminate (Riedel de Haen 13404) was added as an additive to depress the high Si/Al ratio of the raw materials on values around 1.0 found in the technical important zeolites LTA and LSX (the low silica form (Si:Al = 1) of zeolite X).

The amounts of NaAlO$_2$ were therefore adapted to the individual SiO$_2$ and Al$_2$O$_3$ contents of the filtration residues (see Table 1). Always 4 g FR and NaAlO$_2$ amounts of 3.2 g (FR 1) or 4.8 g (FR 2) were mixed and 40 ml NaOH (Fluka 71691) were added as mother liquor of synthesis. The resulting solid/liquid ratio of 0.1 kg/l was selected to prevent production of higher amounts of waste aqueous NaOH solutions. In a first series the influence of the alkalinity was investigated by inserting NaOH solutions of high concentration, calculated according the total caustic ratio "Na$_2$O:H$_2$O" equal to 8 M, 12 M and 16 M NaOH. The main components within the solutions after the reactions were analysed by ICP-OES to develop possible ways for a reinsertion in zeolite formation process. A constant duration of 4 h was revealed for each experiment of the series on the influence of alkalinity.

Table 1: Chemical composition of the filtration residues FR 1 and FR 2

Chemical composition (% by mass)*										
	SiO$_2$	Al$_2$O$_3$	Fe$_2$O$_3$	TiO$_2$	CaO	MgO	Na$_2$O	K$_2$O	SO$_3$	Cl⁻	Loss of ignition
FR1	80.50	2.32	1.30	0.12	0.85	4.25	4.60	0.74	4.84	0.45	14.62
FR2	92.3	3.83	0.49	0.05	2.05	0.19	0.01	0.05	0.23	3.23	3.23

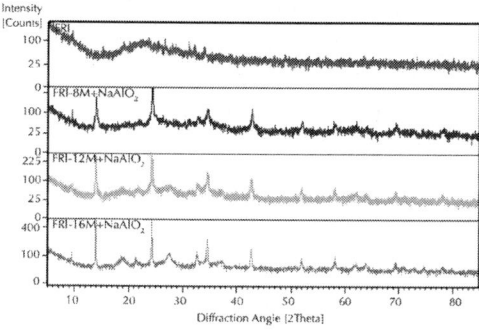

Figure 1: X-ray powder patterns of FR 1 and the products of FR 1 + NaAlO$_2$ according to direct one pot synthesis at 60°C in dependence of the alkalinity.

Figure 2: SEM images of the reaction products according to Table 3: left column: FR 1 raw material (a); FR 1-1h (b); FR 1-3h (c); FR 1-12h (d) and FR 2 right column: FR 2 raw material (e); FR 2-1h (f); FR 2-3h (g); FR 2-12h (h).

Beside variation of alkalinity the one pot procedure was even performed to study the importance of the reaction time between 1 - 3 h. An alkalinity of 8 M NaOH was therefore selected for both FR 1 and FR 2 on the basis of the results of the experiments under variation of the alkalinity. An additional sample with FR 1 as well as FR 2 was synthesized at a time of 12 h for comparison with result at the shorter reaction periods.

The whole experiences from one pot syntheses were then focused to develop a second synthesis pathway as alternative ways for production of pure phase zeolites A and X using FR 1 and FR 2. This "two step reaction route" started with an alkaline leaching of FR in 8 M NaOH at 60°C for 3 h. After diluting the alkaline slurry with water and mixing with NaAlO$_2$ gel crystallization was followed under usual conditions at 353 - 363 K for 3 - 16 h in steel autoclaves (see Table 2).

Independent from the reaction route all products were washed with 150 ml water and dried at 80°C before characterization by diverse analytical methods.

Analytical Methods

Chemical analyses of selected reaction products (solid products and reaction solutions) were performed by inductively coupled plasma optical emission spectrometry (ICP-OES) as already described for the starting materials (chapter 2.1.).

The reaction process under variation of the alkalinity was followed gravimetrically too by weighting the solid products after each experiment on a Kern EMB 200-3 laboratory balance. Thus, first information on the digestion of the raw materials under the conditions of the different alkaline solutions were expected by evaluation of the mass of solids parallel to phase analysis by XRD as well as quantitative chemical analysis of the solutions after each experiment.

All reaction products were examined by X-ray powder diffraction (XRD) using a BRUKER D4 Endeavor powder diffractometer, with Bragg Brentano geometry and Cu Kα radiation at 40 kV and 40 mA. 2668 steps of 0.03° step wide were measured at 1 s measuring time per step in the 2 Theta ranges 5° - 85°. The winXpow software (Stoe & Chi GmbH, Darmstadt) was revealed for data evaluation.

In addition to XRD FTIR spectroscopy was performed using a BRUKER IFS66v Fourier Transform IR spectrometer. KBr pellets were pressed with 200 mg KBr and 1 mg of the sample. According to their IR fingerprint [18] additional structural information on the zeolites were expected from the spectra. A simple detection of possible impurities from their typical vibration modes [19] like hydroxyl groups and carbonate anions, which are introduced within the products by the strong alkaline NaOH solution during the reactions, was a further aim of the FTIR investigation.

Crystal size and morphology of optimal products were characterized by scanning electron microscopy (SEM) on a JEOL JSM-6390A scanning electron microscope at an acceleration voltage of 20 kV, using Au-sputtered samples for optimal charge flow.

The water content of the products was measured by simultaneous thermal analysis on a Setaram Setsys Evolution t1750 thermoanalyzer. Samples were heated in an atmosphere of technical air (80 ml N_2 and

20 ml O_2, flow rate 20 ml/min) up to 600°C at a rate of 5°C/min. After 30 min holding time at 600°C samples were cooled down to room temperature.

Additionally BET surface area of the zeolites was investigated by nitrogen sorption method at 77 K using a Micromeritics Accelerated Surface Area and Porosimetry System ASAP 2010.

RESULTS

Direct One Pot Synthesis at 60°C in Dependence of the Alkalinity

The products of direct one pot synthesis, observed at 60°C in dependence of the alkalinity are given in Table 2. Some additional experimental details and the mass ratio between solid products and solid educts as well as the insertion of SiO_2 into the solid products are added to this table. The latter values were calculated according to the results calculated free of loss on ignition of chemical analysis, of solid FR 1 and 2 (Table 1) as well as analyses of the solutions, summarized in Table 3. Phase analysis was performed according to qualitative XRD analyses. The X-ray powder patterns of all these products are given in Figure 1, Figure 3.

Table 2: Experimental conditions, syntheses products, mass ratio of solids (product/educts) and insertion of SiO_2 into solid products of direct one pot syntheses in dependence of the alkalinity (temperature 60°C, duration 4 h)

No	Sample	NaOH Mol/l*	NaAlO$_2$(g)	Products**	Mass ratio product/ educt	SiO$_2$ in solid product (%)
FR 1						
1	FR1-8M+NaAlO$_2$	8	3.2	INT-1 + A	0.85	66
2	FR1-12M+NaAlO$_2$	12	3.2	INT-2 + A	0.78	57
3	FR1-16M+NaAlO$_2$	16	3.2	INT-2 + (A)	0.80	53
FR 2						

4	FR2-8M+NaAlO$_2$	8	4.8	SOD + (Z-21)	0.94	76
5	FR2-12M+NaAlO$_2$	12	4.8	SOD + Z-21	0.95	69
6	FR2-16M+NaAlO$_2$	16	4.8	SOD	0.91	71

*Molarity as calculated from the inserted NaOH + Na$_2$O content of the additive NaAlO$_2$; **INT-1: intermediate zeolite with random disorder between sodalite and cancrinite [20] [21] ; INT-2: intermediate zeolite with transition of stacking sequence to cancrinite [20] [21] ; A: amorphous aluminosilicate; SOD: hydrosodalite [28] [29] ; Z-21: zeolite Z-21 (PDF 27-1405 [17] [30]); ( ): low amounts.

Table 3: Main components of the reaction solutions of FR 1 and FR 2 from of direct one pot syntheses in dependence of the alkalinity at a temperature of 60°C and a duration of 4 h

Ex.No.	NaOH(M)	Chemical composition of reaction solution									
		SiO$_2$ (g/l)	Al$_2$O$_3$ (g/l)	Fe$_2$O$_3$ (mg/l)	TiO$_2$(mg/l)	CaO (mg/l)	MgO (mg/l)	Na$_2$O (g/l))	K$_2$O (g/l)	Cl$^-$ (mg/ml)	SO$_3$(mg/ml)
Alkaline solution of reaction of FR 1 + NaAlO$_2$											
1	8	27.19	0.37	14.34	1.10	2.04	0.37	128.2	1.46	142.88	120.70
2	12	34.18	0.88	34.46	0.80	2.42	0.78	202.45	1621.0	198.95	45.95
3	16	38.01	1.91	94.39	12.57	0.08	3.34	256.70	1644.0	256.55	3.40
Alkaline solution of reaction of FR 2 + NaAlO$_2$											
4	8	21.86	0.58	35.05	0.49	1.02	0.76	128.95	0.14	181.47	3.35
5	12	28.01	1.57	85.12	1.33	1.86	0.84	205.30	224.2	192.36	1.40
6	16	26.78	3.64	35.48	2.49	4.90	1.19	258.05	315.05	256.55	8.74

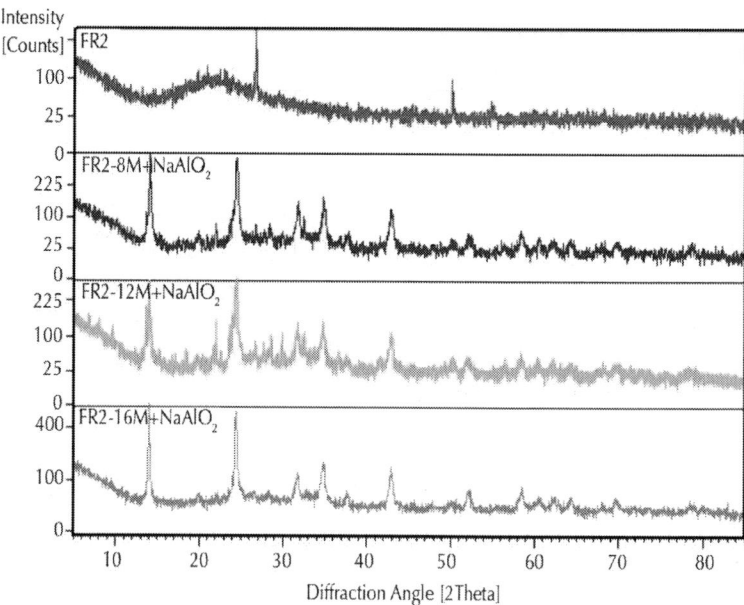

Figure 3: X-ray powder patterns of FR 2 and the products of FR 2 + NaAlO$_2$ according to direct one pot synthesis at 60°C in dependence of the alkalinity.

From Table 2 and Figure 1, Figure 2 it can be derived that the experiments with FR 1 and FR 2 yield to somewhat different reaction products under the applied low temperature (60°C) conditions and 4 h duration. Only zeolite INT, a phase with intermediate structure between sodalite (SOD) and cancrinite (CAN) [20] [21] was obtained as crystalline reaction product from FR 1 beside amorphous aluminosilicate in 8 M NaOH. The sodalite framework is formed by ABC stacking of layers of not directly connected (AlSiO$_4$)$_3$ - rings along [111] as in cubic closest package. The cancrinite structure exhibits the AB stacking of the same layers parallel [0001] known from hexagonal closest package [22] -[27] . Random stacking between both sequences results in the structure of INT and as a consequence of this disorder the X-ray powder pattern of the intermediate phase shows only the reflections that sodalite and cancrinite have in common [20] [21] as can be seen in the patterns of Figure 1. At higher alkalinities (exp. No. 2 and 3, Table 2 and products obtained with 12 M and 16 M NaOH in Figure 1) even zeolite INT was formed exclusively. But the evaluation of two further broad peaks at 19° 2θ and 28° 2θ in the pattern indicates slight transition from random disorder to more ordered structure and the product is thus termed INT-2 in Table 2. As both broad signals are the main lines of cancrinite, structure shows more and more features in direction of a (disordered) cancrinite. As carbonate anions have a strong structure directed effect on cancrinite formation the higher content of carbonate impurities of the super alkaline conditions of exp. No. 2 and 3 (Table 2) is responsible for this structural shift.

In contrast, sodalite is formed in each case the main phase under insertion of FR 2. This phase was proved to be an expanded hydrosodalite $Na_6[AlSiO_4]_6 \cdot (H_2O)_4$ (PDF 42-216, [17] and [28] [29]). From Figure 4 formation of a minor second phase can be observed by insertion of FR 2. Here zeolite Z-21 (PDF 27-1405 [17] [30]) occurs during reaction of FR 2 under alkalinities of 8 M and 12 M NaOH (see Figure 3 and exp. No. 4-5, Table 2) whereas pure phase hydrosodalite again exists under strongest alkaline conditions of 16 M NaOH as single phase (exp. No. 6, Table 1).

From the mass ratios, given in Table 2 no remarkable influence of the alkalinity on the course of the reactions can be derived. The values are similar to one and the same raw material but differif FR 1 or FR 2 is inserted. Whereas a lower range between 0.78 - 0.84 was observed by using FR 1 the interval 0.91 - 0.95 could be found for FR 2. In comparison with FR 2, the lower ratios of the reaction products with FR 1 seem to be more a result of the different composition of the two filtration residues (see Table 1) than of remarkable differences in the course of the alkaline reactions. Thus, the higher silicate content of FR 2 mainly reflects these ratios. Most of this amount seems to be incorporated within the solid products whereas components like MgO and SO_3 enriched in FR 1 will not enter the solid products if exclusively sodium aluminosilicate hydrates were formed during the reactions.

Beside this information from mass ratios, the distribution of SiO_2 between the solid products and the solution gives some further information on the course of the reactions. Table 3 summarizes the data of chemical analysis of the reaction solutions from direct one pot syntheses experiments in dependence of the alkalinity. The insertion of SiO_2 into the solid products (see Table 2) was calculated according to chemical analysis data of the solid FR 1 and FR 2 (free of loss of ignition, see Table 1) as well as the analyses of the solutions, summarized in Table 3. In general the incorporation of SiO_2 into the solid phase is somewhat higher in the reactions with FR 2. Independent from the raw materials the highest conversion can be stated for reactions in 8 M NaOH for both (66% for FR 1 and 76% for FR 2 (see Table 2). About a third (FR 1) and a quarter (FR 2) of the silicate content will thus remain in the solutions under these conditions (see Table 3) and a re-use of the solutions should be tested in forthcoming experiments.

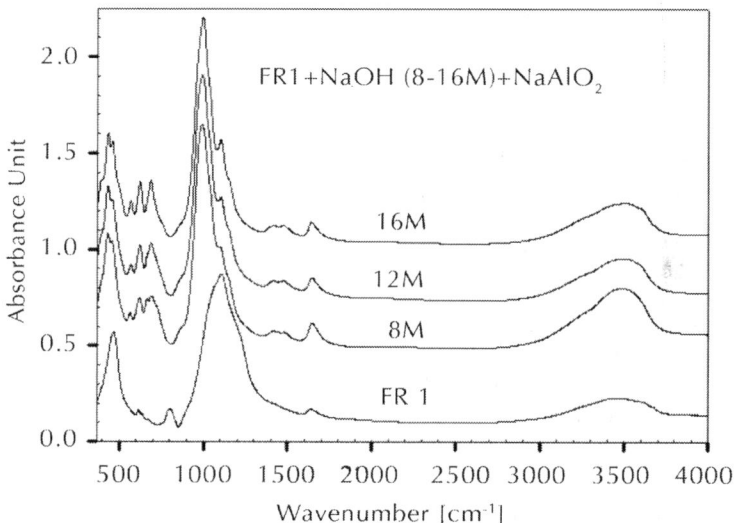

Figure 4: FTIR spectra of FR 1 and its products of direct one pot synthesis at 60°C in dependence of the alkalinity.

The FTIR spectra of reaction products of FR 1 + NaAlO$_2$ in 8 M, 12 M and 16 M NaOH (No. 1-3,Table 2), recorded in the 400 - 4000 cm^{-1} range, are summarized in Figure 4. The spectrum of the raw material FR 1 is included in Figure 4 for comparison. As the main component of FR 1 is amorphous silica its spectrum exhibits weak resolved modes at wave-numbers 458 cm^{-1}, 506 cm^{-1}, 692 cm^{-1}, 778-797 cm^{-1}, 1078 cm^{-1} and 1164 cm^{-1} all around the region, known from quartz [31] .

The FTIR spectrum "8 M" in Figure 4 (product, obtained in 8 M NaOH according exp. No. 1, Table 2) shows close resemblance with cancrinite (CAN) [18] [20] [21] [31] [32] . XRD revealed this product as intermediate phase between sodalite and cancrinite (INT). As in cancrinite, the intense broad asymmetric T-O-T (T = Si, Al) stretching mode at ~1000 cm^{-1} and a sharp weaker one around 1100 cm^{-1} are well resolved. In the range of the symmetric T-O-T stretching vibrations and double ring modes of cancrinite (755 cm^{-1} - 567 cm^{-1} according to Flanigen et al. [18]) the 755 cm^{-1} mode is absent in the spectrum of INT. Furthermore the bending modes at ~ 460 cm^{-1} and 430 cm^{-1} were observed as in cancrinite, whereas the mode around 500 cm^{-1} is not resolved in the spectrum of INT but found in cancrinite [21] [32]. In addition a band at about 730 cm^{-1} can be observed in the spectrum of the intermediate phase not found in cancrinite but in sodalite [18] , a hint of deviation in random stacking disorder in favour of the sodalite sequence [20] [21] . Thus, the deviation in the spectra of the intermediate phase and cancrinite are clearly reflecting the structural characteristics of both phases as result of the stacking disorder of INT, as mentioned above. The vibrations of the water molecules as well as of carbonate enclathrated within the cages of the intermediate phase, were also observed in the spectrum according to the characteristic IR active bands of water (bending mode at 1650 cm^{-1} and O-H vibrations at 3100 - 3600 cm^{-1}) and the asymmetric stretching mode of the carbonate group as doublet (1410 - 1450 cm^{-1} [19]).

The FTIR spectra "12 M" and "16 M" in Figure 4 of the products according exp. No. 2 and 3, Table 2, both exhibit very close resemblance with the spectrum of the 8 M synthesis. In accordance with XRD, as also the intermediate zeolite between sodalite and cancrinite has been formed in both experiments. As a sign of a shift of stacking disorder in favour of cancrinte, the band at about 730 cm^{-1} (sodalite) is absent now and all bands are sharper, as can be seen at the asymmetric T-O-T stretching mode at 1100 cm^{-1} especially in the "16 M" product (see Figure 4).

The FTIR spectra of reaction products of FR 2 + NaAlO$_2$ in 8 M, 12 M and 16 M NaOH (No. 4-6, Table 2) are summarized in Figure 5. The spectrum of the raw material FR 2 is included in this figure for comparison and has close resemblance with the spectrum of FR 1. As identified in FR 1, the bands of SiO$_2$ can be seen again. Due to the amorphous character of FR 2, the bands are broad and not well resolved, especially the weaker quartz vibrations at 692 cm^{-1}, 778 - 797 cm^{-1} [31].

The FTIR spectrum of the product obtained in 8 M NaOH in Figure 5 (according exp. No. 4, Table 2) exhibits the vibrations of sodalite host framework in agreement with literature data [18] [31] . The strong asymmetric T-O-T (T= Si, Al) stretching vibration at 1000 cm^{-1} and the weaker triplet of symmetric T-O-T vibrations in the 660 cm^{-1} - 740 cm^{-1} beside the bending modes in the 400 cm^{-1} - 500 cm^{-1}region are well resolved in this spectrum. The water molecules of the hydrosodalite (bending mode at 1650 cm^{-1}; O-H-vibrations as broad band at 3100 cm^{-1} - 3600 cm^{-1} and carbonate impurities at 1410 cm^{-1} and 1450 cm^{-1} [19]) can be clearly seen in the spectrum.

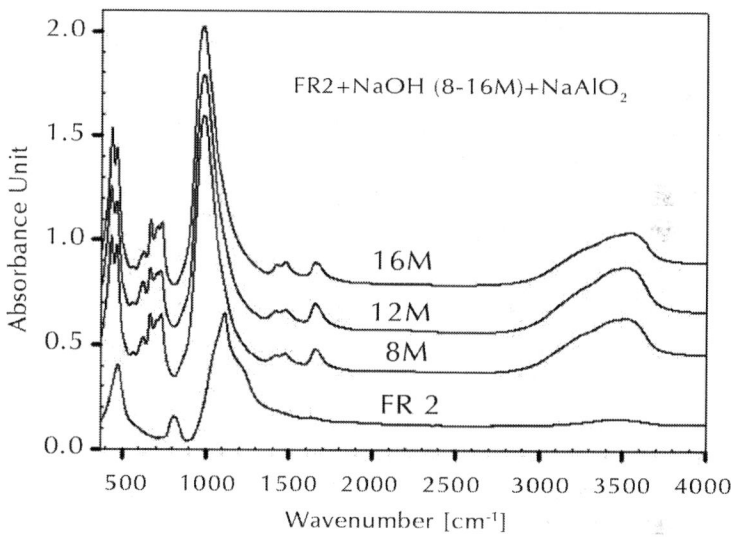

Figure 5: FTIR spectra of FR 2 and its products of direct one pot synthesis at 60°C in dependence of the alkalinity.

A further band at 630 cm^{-1} is caused by the amorphous aluminosilicate parts of the product according to Al-O-Si stretching vibrations as known from Al-Si-geopolymers [33]. Compared with products of FR 2, all products from FR 1 also contain amorphous parts but the 630 cm^{-1} vibration is overlapped here by strong absorption mode of the intermediate phase, formed with FR 1.

The FTIR spectra of the two products FR 2 "12 M" and "16 M" in Figure 5 (according exp. No. 5 and 6, Table 2) show very close resemblance with the spectrum of the "8 M" product, as hydrosodalite was the main product in "12 M" synthesis and the single product under conditions of "16 M" NaOH. The few amounts of zeolite Z 21 in the

"12 M" product cannot be distinguished by further bands in the FTIR spectrum, but differences in signal intensities and less pronounced signals in the region of aluminosilicate framework vibrations seem to be a result of the two phase mixture.

Direct One Pot Synthesis at 60°C in Dependence of the Reaction Time

The clarification that zeolites can successfully synthesized at low temperatures at 60°C under strong alkaline conditions was first given by Hadan and Fischer [14] and Fischer et al. [15] [16] . The authors used gels from pure chemicals. During their experimental study the important role of the reaction time under hyperalkaline conditions was further pointed out. If zeolites always crystallize under kinetic control, a time dependent crossover of zeolites Na-A and sodalite seems possible already in the period from 1 to 4 h duration [14].

In contrast to the results presented in literature, the experiments of direct one pot synthesis at 60°C are performed with waste materials FR 1 and FR 2 instead of pure chemical sodium silicate. Microporous zeolites INT or SOD is formed at a constant period of 4 h (see passage 3.1 above). Thus, testing the influence of the reaction time is now the second step of the experimental investigation. The early period of reactions at 1 h, 2 h and 3 h and synthesis at prolonged time (12 h) were therefore examined. As the experiments of one pot synthesis under variation of the alkalinity yield microporous zeolites already in 8 M NaOH and no other phases up to 16 M NaOH (see former chapter) the time dependent reactions were performed in mother liquor equivalent to 8 M NaOH. To do not insert strong alkaline solutions, it is also meaningful for future applications in industrial waste recovery process from an economical point of view. Furthermore the production of high quantities of hyperalkaline waste solutions is prefented. The experimental conditions are given in Table 4.

The X-ray powder patterns of the products of time dependant reactions with FR 1 and FR 2 are summarized in Figure 6 and Figure 7. SEM images of these samples are shown in Figure 6. From XRD of the FR 1 series, formation of the intermediate zeolite between sodalite and cancrinite, termed INT-1 (Table 3) can be observed already after 1 h reaction time beside amorphous contributions with no changes even after 2 hours. Whereas the SEM image of untreated FR 1 raw material indicates very small nanoparticles (Figure 2(a)), the 1 h product contains the intermediate zeolite as spheroidal crystals of remarkable size between 0.5 - 0.8 µm (see Figure 2(b)).

Table 4: Experimental conditions and syntheses products of direct one pot syntheses in dependence of the reaction time (temperature 60°C and 8 M NaOH for both series)

No.	sample	Reaction time (h)	NaAlO$_2$ (g)	Products[*]
FR 1				
7	FR1-8M+NaAlO$_2$-1h	1.0	3.2	INT-1 + A
8	FR1-8M+NaAlO$_2$-2h	2.0	3.2	INT-1 + A
9	FR1-8M+NaAlO$_2$-3h	3.0	3.2	INT-1 + U+ A
10	FR1-8M+NaAlO$_2$-12h	12.0	3.2	INT-3 + (A)
FR 2				
11	FR1-8M+NaAlO$_2$-1h	1.0	4.8	NaA + SOD + A
12	FR1-8M+NaAlO$_2$-2h	2.0	4.8	SOD + NaA + A
13	FR1-8M+NaAlO$_2$-3h	3.0	4.8	SOD + A
14	FR1-8M+NaAlO$_2$-12h	12.0	4.8	SOD + (A)

[*]INT-1: intermediate zeolite with random disorder between sodalite and cancrinite [20] [21] ; A: amorphous aluminosilicate; SOD: hydrosodalite [28] [29] ; INT-3: intermediate zeolite with transition of stacking sequence to sodalite [20] [21] Z-21: zeolite Z-21 (PDF 27-1405 [17] [30]); U: unknown phase with some resemblance to sodium aluminum silicate hydrate PDF 43-577 [17] ; ( ): low amounts.

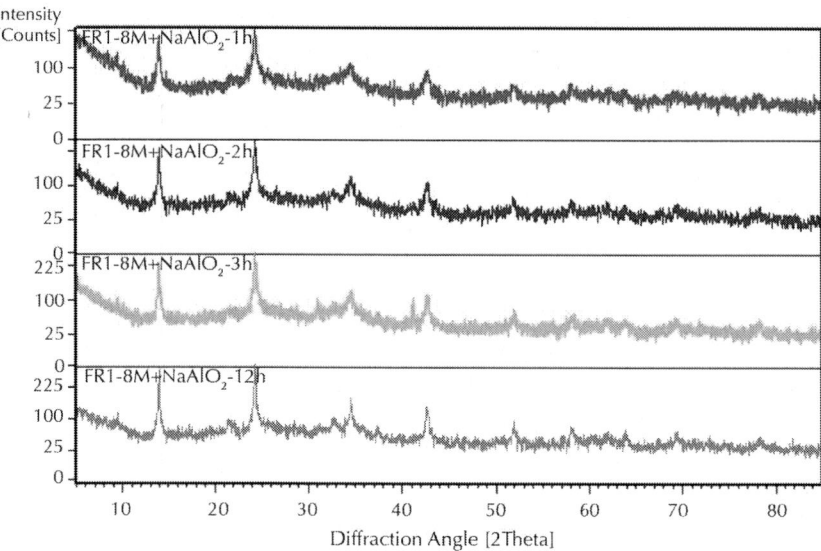

Figure 6: X-ray powder patterns of products obtained from FR 1 + NaAlO$_2$in direct one pot synthesis at 60°C in dependence of the reaction time.

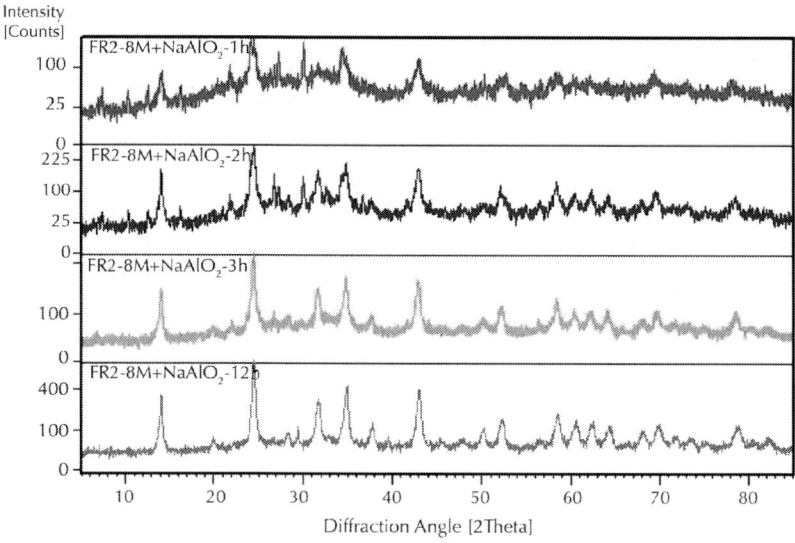

Figure 7: X-ray powder patterns of products obtained from FR 2 + NaAlO$_2$in direct one pot synthesis at 60°C in dependence of the reaction time.

After the reaction period of three hours the same phase is obtained but it contains now surprisingly much smaller crystals (see Figure 6 and Figure 2(c)). Some new but weak reflections appear (for instance at 31° and 41° 2 θ) beside INT in the powder pattern of the 3 h product (Figure 2(c)). The new signals belong to a second phase (U, Table 4) that cannot definitely be identified due to the contribution of the strong pattern background. As this phase already disappears after 4 h (seeTable 2 and Figure 1), its field of stability is located in a narrow window between 2 h - 4 h reaction time.

In the 12 h product the intermediate zeolite between sodalite and cancrinite still exists as single phase. An appearance of a weak broad 530 reflex at 59° 2 θ and a 200 reflex at 20° 2 θ gives now a hint on the more sodalite like character of this phase termed INT-3 in Table 4 [20] [21] . This structural change can also be followed in Figure 2(d), where a shift from spherical to more polyhedral morphology can be seen.

The XRD patterns of the time dependant series with FR 2 are summarized in Figure 7. SEM images of selected products are given in Figures 2(e)-(h). A cocrystallization of remarkable amounts of zeolite NaA and hydrosodalite are found after 1 h reaction time. From the very fine grained FR 2 (see SEM picture Figure 2(e)) formation of big cube-like crystals of NaA beside spherulitic sodalites can be stated from the SEM photograph given in Figure 2(f). At prolonged times of 2 h and 3 h the amount of zeolite NaA is decreasing continuously (see Figure 7 and Figure 2(g)) and only a few cubes of zeolite A can be found during the SEM investigation (Figure 2(g)).

After 12 h hydrosodalite exists as single phase with high degree of crystallinity, as shown by the remarkable increase of intensities and the decrease of the background in the XRD pattern. Hydrosodalite crystals of around 1 μm with spherical morphology were detected during SEM investigation within the 12 h product (see Figure 2(h)).

Two Step Synthesis of Zeolites Na-A and Na-X from the Raw Materials FR 1 and FR 2

In the one pot direct syntheses under hyperalkaline conditions at 60°C the technical important zeolites of high commercial interest like NaA could not be observed as pure phase by insertion of the filtration residues. Only a cocrystallisate of zeolite NaA and hydrosodalite could be synthesized from FR 2 as the only sample with some potential technical importance. The other products like pure hydrosodalite or the intermediate zeolite between sodalite and cancrinite have no technical application up to now. Nevertheless the direct syntheses experiments yield successful new information on the principle behaviour of the waste material FR under low temperature alkaline conditions and addition of $NaAlO_2$.

With help of these findings on the digestion and crystallization behaviour of the FR-s under hyperalkaline low temperature conditions, a second synthesis pathway is now developed for transformation of FR 1 and 2 into zeolites NaA and NaX. This reaction route, a "two step synthesis" first started with an alkaline leaching of FR in 8 M NaOH at 60°C for 3 h. After the alkaline slurry was diluted with water and mixed with $NaAlO_2$ gel crystallization was followed under usual conditions at 353 - 363 K for 3 - 16 h in Teflon coated steel autoclaves. The amounts between 2 g and 0.5 g $NaAlO_2$ were inserted to test even formation of zeolite NaX in the silicate rich synthesis batch. The experimental conditions and syntheses products are summarized in Table 5 and the results of X-ray powder diffraction are summarized in Figure 8 and SEM photographs are shown in Figure 9. From Table 5 and the Figure 8, Figure 9 formation of pure phase zeolites NaA (No. 13, Table 5) and a co-crystallization of NaX and INT-1 (No. 14, Table 5) could be derived from these fine-tuned syntheses conditions under insertion of FR 1. SEM analysis of the products of FR 1 show cubes of 1 μm in size for zeolite NaA and intergrowths of NaX and INT-1 of sizes up to 1.5 μm (see Figure 9, products FR 1-A and FR 1-X, photos on top).

Use of FR 2 yield to intergrowth of NaA cubes and octahedral crystals of NaX of size <1 μm in exp. No. 15 and pure phase zeolite NaX of octahedral shape and crystal size around 2 μm in exp. No. 16, Table 4 (see Figure 9, products FR 2-A and FR 2-X, bottom photos).

The FTIR spectra of the zeolites NaA (FR 1-A and FR 2-A) and NaX (FR 1-X and FR 2-X), obtained from the two step syntheses according Table 5 are summarized in Figure 10. The spectra of commercial zeolites NaA (A-Std) and NaX (X-Std) are included in this figure as standard samples for comparison. The typical vibrations of zeolite A and X framework structures can be clearly seen for the pure phase products FR 1-A and FR 2-X in Figure 10, obtained in experiments No.

13 and 16, Table 5. The spectra are similar to the commercial zeolite standards and the literature data [18]. In agreement with XRD data, the two phase character of product No. 14 (zeolite Na A + INT-1) and No. 15 (mixture of Na A and some NaX) were observed even in the FTIR spectra of these samples. Thus, spectrum of product No. 14, INT-1 shows the double ring mode around 624 cm^{-1} and the symmetric T-O-T stretching mode at 680 cm^{-1}. The other bands are overlapped by the strong vibration modes of the NaX-framework constituents. The small contribution of zeolite NaX in sample No. 15 (Figure 10 and Table 5) can be clearly observed according to the symmetric T-O-T stretching mode at 746 cm^{-1} in agreement with literature data [18].

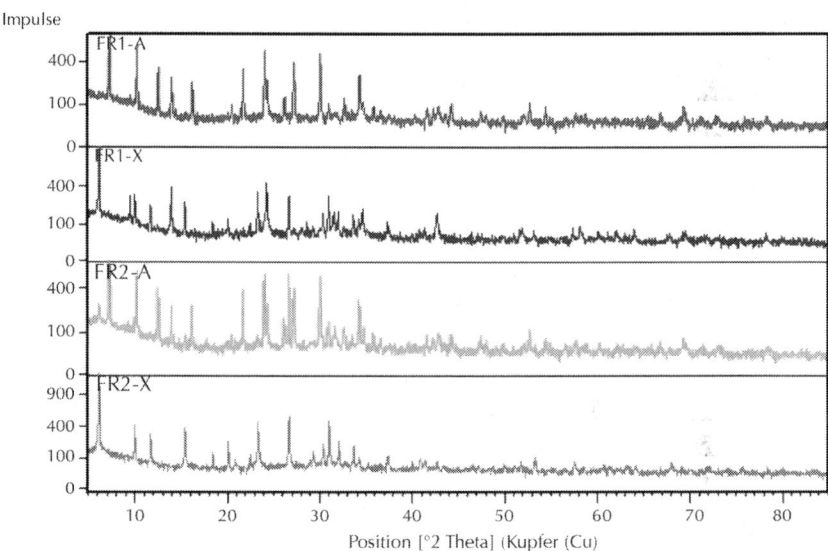

Figure 8: X-ray powder patterns of the products obtained by two step syntheses with FR 1 and FR 2.

Figure 9: SEM micrographs of the zeolites NaA (FR 1-A and FR 2-A) and NaX (FR 1-X and FR 2-X), obtained from the two step syntheses according Table 5.

The water molecules of the zeolites were observed in all spectra due to the bending mode of H_2O at 1650 cm^{-1} and the broad band of O-H-vibrations between 3100 cm^{-1} - 3600 cm^{-1}. Some carbonate impurities were detected only in the product F R2-X according to the characteristic asymmetric stretching mode of the carbonate group in the 1410 cm^{-1} - 1450 cm^{-1} [19]) region.

For a further characterization of the zeolite products, obtained from FR 1 and FR 2 by the two step synthesis route the BET surface area and the water content was measured. The data are summarized in Table 6.

Sufficient physicochemical properties can be expected from these values for the products FR 1-A and FR 2-X as compared with literature data of commercial zeolites [2] . As the pore system of zeolite A is not accessible for nitrogen molecules due to the narrow 4 Å opening windows of NaA structure, the low BET-value results for FR1-A. The slightly higher value of FR 2-A is caused by the small portions of zeolite X within the product (see Table 6). The deviation between the X zeolites has similar reasons: compared with FR 2-X the BET-surface area of FR 1-X is much too low, caused by the cocrystallization with zeolite INT-1 (see Table 6).

Table 5: Experimental conditions and syntheses products of the two step syntheses

No.	sample	Step No., solid educts and solvent	Temp. (°C)	time (h)	Products* (Wt %)
13	FR1-A	1. FR 1 (2 g), NaOH (20 ml, 8 M) 2. Slurry 1+ H_2O (30 ml) +NaAlO$_2$ (2g)) + 20 ml H_2O	60 80	3 22	Alkaline FR 1-slurry NaA
14	FR1-X	1. FR 1 (2g), NaOH (20 ml, 8 M) 2. Slurry 1+ H_2O (30 ml) +NaAlO$_2$ (1.2 g)) + 20 ml H_2O	60 90	3 16	Alkaline FR 1-slurry NaX + INT-1
15	FR2-A	1. FR 2 (2g), NaOH (20 ml, 8 M) 2. Slurry 1+ H_2O (20 ml) +NaAlO$_2$ (1.6 g) + 10 ml H_2O	60 90	3 16	Alkaline FR 2-slurry LTA NaA + NaX
16	FR2-X	1. FR 2 (2g), NaOH (20 ml, 8 M) 2. Slurry 1+ H_2O (30 ml) + NaAlO$_2$ (0.5 g) + 20 ml H_2O	60 80	3 22	Alkaline FR 2-slurry NaX

*INT-1: intermediate zeolite between sodalite and cancrinite; SOD: hydrosodalite; NaA: zeolite NaA; NaX: zeolite NaX.

Table 6: BETsurface area and water content of the zeolites NaA (FR1-A and FR 2-A) and NaX (FR 1-X and FR 2-X), obtained from two step syntheses according Table 5

sample	Exp. No. (acc. Tab. 6)	BET-surface area [m²/g]	water content [%]	H_2O-molecules per formula unit
FR1-A	13	49.4 (1)	16.25	20
FR1-X	14	353.2 (2)	16.94	171
FR2-A	15	102.0 (1)	17.44	21
FR2-X	16	644.6 (1)	21.09	213

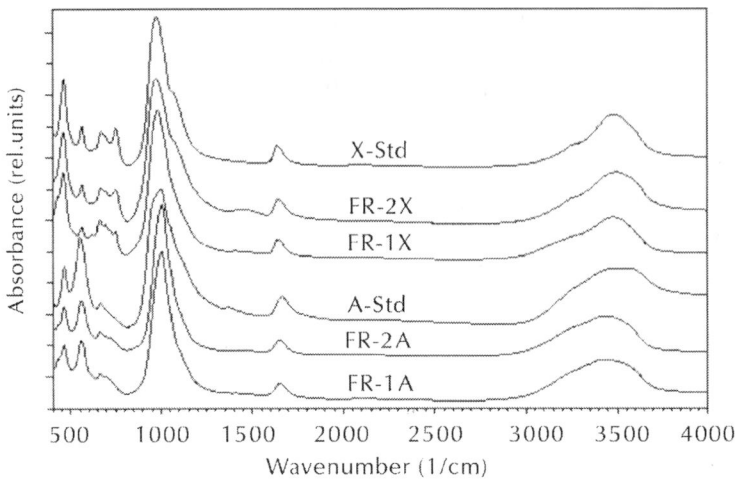

Figure 10: FTIR spectra of the zeolites NaA (FR 1-A and FR 2-A) and NaX (FR 1-X and FR 2-X), obtained from the two step syntheses accordingTable 5. Spectra of commercial zeolites NaA (A-Std) and NaX (X-Std) are included as standard samples for comparison.

CONCLUSIONS

Principles of zeolite formation under use of filtration residues FR under strong alkaline conditions at low reaction temperatures could be shown from our experimental results. Compared with literature data of synthesis with gels obtained from pure chemicals, a remarkable shift of the onset of the zeolite NaA/hydrosodalite crossover reaction could be stated. Thus by the use of FR in direct one pot synthesis at 60°C a pure phase zeolite NaA was not observed in the 1 h - 12 h interval. An alkalinity equivalent to insertion of 8 M NaOH was tested to be fully suitable for a rapid dissolution of FR and insertion of nucleation of zeolite phases already within the first hour of reaction.

The deviation in composition of the two filtration residues is expected to be responsible for the differences of the reaction under kinetic control and affects the structure type of the products. Either formation of zeolite INT with FR 1 (containing 80% by mass SiO_2) or hydrosodalite with FR 2 (containing 93% by mass SiO_2) were observed. Zeolite INT could be of interest in industry too, because the stacking disorder between the structures of sodalite and cancrinite will block the formation of a main structural channel. Cages of different size and length are the result. Thus, modified zeolitic behaviour of INT is expected.

Crystallization of pure phase NaA or NaX zeolites from both filtration residues is possible by alkaline low temperature digestion in 8 M NaOH for 3 h followed by a second crystallization step under addition of sodium aluminate and conditions of 80°C - 90°C and times between 16 - 22 h.

Further fine tuning of reaction conditions of one pot synthesis as well as possible re-use of the reaction solutions should be tested in forthcoming experiments.

ACKNOWLEDGEMENTS

We would like to thank the Deutsche Forschungsgemeinschaft for funding this research project.

REFERENCES

1.   Barrer, R.M. (1978) Zeolites and Clay Minerals as Sorbents and Molecular Sieves. Academic Press, London.

2.   Breck, D.W. (1984) Zeolite Molecular Sieves: Structure, Chemistry and Use. John Wiley & Sons Inc., New York.

3.   Barrer, R.M. (1982) Hydrothermal Chemistry of Zeolites. Academic Press, London.

4.   Aiello, R., Colella, C., Casey, D.G. and Sand, L.B. (1980) Experimental Zeolite Crystallization in Rhyolitic Ash-Sodium Salt Systems. Proceedings of 5th International Zeolite Conference, London, 1980, 49-55.

5.   Höller, H. and Wirsching, U. (1985) Zeolite Formation from Fly Ash. Fortschr. Miner, 63, 21-43.

6.   Grutzeck, M. and Siemer, D.D. (1997) Zeolites Synthesized from Class F Fly Ash and Sodium Aluminate Slurry. Journal of the American Ceramic Society, 80, 2449-2453.http://dx.doi.org/10.1111/j.1151-2916.1997.tb03143.x

7.   Maenami, H., Shin, H., Ishida, H. and Mitsuda, T. (2000) Hydrothermal Solidification of Wastes with Formation of Zeolites. Journal of Materials in Civil Engineering, 12, 302-306.http://dx.doi.org/10.1061/(ASCE)0899-1561(2000)12:4(302)

8.   Miyake, M., Tamura, C. and Matsuda, M. (2002) Resource Recovery of Waste Incinerator fly Ash: Synthesis of Zeolites A and P. Journal of Materials in Civil Engineering, 85, 1873-1875. http://dx.doi.org/10.1111/j.1151-2916.2002.tb00368.x

9.   Murayama, N., Yamamoto, H. and Shibata, J. (2002) Zeolite Synthesis from Coal Fly Ash by Hydrothermal Reaction Using Various Alkali Sources. Journal of Chemical Technology and Biotechnology, 77, 280-286. http://dx.doi.org/10.1002/jctb.604

10. Anuwattana, R. and Khummongkol, P. (2009) Conventional Hydrothermal Synthesis of Na-A Zeolite from Cupola Slag and Aluminium Sludge. Journal of Hazardous Materials, 166, 227-232. http://dx.doi.org/10.1016/j.jhazmat.2008.11.020

11. Di Renzo, F., Fajula, F., Figueras, F., Nicolas, S. and Des Courieres, T. (1989) Are the General Laws of Crystal Growth Applicable to Zeolite Synthesis? In: Jacobs, P.A. and van Santen, R.A., Eds., Zeolites: Facts, Figures, Future, Studies in Surface Science and Catalysis, Vol. 49, Part A, Elsevier, Amsterdam, 119-132.

12. Daniels, R.H., Kerr, G.T. and Rollmann, L.D. (1978) Cationic Polymers as Templates in Zeolite Crystallization. Journal of the American Chemical Society, 100, 3097-3100.http://dx.doi.org/10.1021/ja00478a024

13. Weigel, S.J., Gabriel, J.C., Puebla, E.G., Bravo, A.M., Henson, N.J., Bull, L.M. and Sheetham, A.K. (1996) StructureDirecting Effects in Zeolite Synthesis: A Single-Crystal X-ray Diffraction, Si-29 MAS NMR, and Computational Study of the Competitive Formation of Siliceous Ferrierite and Dodecasil-3C (ZSM-39). Journal of the American Chemical Society, 118, 2427-2435. http://dx.doi.org/10.1021/ja952418g

14. Hadan, M. and Fischer, F. (1992) Synthesis of Fine Grained NaA-Type Zeolites from Superalkaline Solutions. Crystal Research and Technology, 27, 343-350.http://dx.doi.org/10.1002/crat.2170270310

15. Fischer, F., Hadan, M. and Fiedrich, G. (1992) Zeolite Syntheses from Superalkaline Reaction Mixtures. Collection of

Czechoslovak Chemical Communications, 57, 788-793.http://dx.doi.org/10.1135/cccc19920788

16. Fischer, F., Hadan, M. and Horn, A. (1991) Investigations to the Synthesis of zeolite Na A for Using in Detergents from Superalkaline Solutions. Chem. Tech, 43, 191-195.

17. JCPDS (1997-2004) International Centre for Diffraction Data, 12 Campus Boulevard, Newton Square.

18. Flanigen, Khatami, H. and Szymanski, H.A. (1971) Infrared Structural Studies of Zeolite Frameworks. Advances in Chemistry, 101, 201-228. http://dx.doi.org/10.1021/ba-1971-0101.ch016

19. Weidlein, J., Müller, U. and Dehnicke, K (1981) Schwingungsfrequenzen. G. Thieme Verlag, Stuttgart.

20. G. Hermeler, J.-Ch. Buhl and W. Hoffmann (1991) The Influence of Carbonate on the Synthesis of an Intermediate Phase between Sodalite and Cancrinite. Catalysis Today, 8, 415-426. http://dx.doi.org/10.1016/0920-5861(91)87020-N

21. Grader, C. and Buhl, J.-C. (2013) the Intermediate Phase between Sodalite and Cancrinite: Synthesis of Nano-Crystals in the Presence of Na_2CO_3/TEA and Its Thermaland Hydrothermal Stability. Microporous and Mesoporous Materials, 171, 110-117. http://dx.doi.org/10.1016/j.micromeso.2012.12.023

22. Gossner, B. and Mussgnug, F. (1930) **Über** Davyn und seine Beziehung zu Hauyn und Cancrinit. Zeitschrift für Kristallographie, 73, 52-60.

23. Baerlocher, C., Meier, W.M. and Olson, D.H. (2001) Atlas of Zeolite Framework Types. 5th Edition, Elsevier, Amsterdam.

24. Jarchow, O. (1962) Zur Struktur des Cancrinits. Fortschr. Mineralogie, 40, 55-56.

25. Jarchow, O. (1965) Atomanordnung und Strukturverfeinerung von Cancrinit. Zeitschrift für Kristallographie, 122, 407-422. http://dx.doi.org/10.1524/zkri.1965.122.5-6.407

26. Pauling, L. (1930) the Structure of Sodalite and Helvite. Zeitschrift für Kristallographie, 74, 213-225.

27. Löns, J. and Schulz, H. (1967) Strukturverfeinerung von Sodalith $Na_8Si_6Al_6O_{24}Cl_2$. Acta Crystallographica, 23, 434- 436. http://dx.doi.org/10.1107/S0365110X67002920

28. Felsche, J. and Luger, S. (1987) Phases and Thermal-Decomposition Characteristics of Hydro-Sodalites Na_8 $[AlSiO_4]_6(OH)_x \cdot nH_2O$. Thermochimica Acta, 118, 35-55.http://dx.doi.org/10.1016/0040-6031(87)80069-2

29. Engelhardt, G., Felsche, J. and Sieger, P. (1992) The Hydrosodalite System $Na_{6+x}[SiAlO_4]_6(OH)_x \cdot nH_2O$ Formation, Phase Composition and Dehydration and Rehydration Studied by 1H, 23N and 29Si MAS-NMR Spectroscopy in Tandem with Thermal Analysis X-Ray Diffraction and IR Spectroscopy. Journal of the American Chemical Society, 114, 1173-1182. http://dx.doi.org/10.1021/ja00030a008

30. Barrer, R.M. and Beaumont, R. (1974) Characterization of the Synthetic Zeolite (Na, Me_4N)-V. Journal of the Chemical Society, Dalton Transactions, 4, 405-407.http://dx.doi.org/10.1039/dt9740000405

31. Moenke, H. (1966) Mineralspektren. Akademie Verlag, Berlin.

32. Buhl, J.-C. (1991) Synthesis and Characterization of the Basic and Non-Basic Members of the Cancrinite-Natrodavyne Family. Thermochimica Acta, 178, 19-31.http://dx.doi.org/10.1016/0040-6031(91)80294-S

Optimal Aerations in the Inverse Fluidized Bed Biofilm Reactor When Used in Treatment of Industrial Wastewaters of Various Strength

Włodzimierz Sokół

Department of Chemical and Bioprocess Engineering, University of Technology and Life Sciences, Bydgoszcz, Poland

ABSTRACT

The aim of this work was the determination of the optimal aerations, and more specifically the corresponding optimal air velocities u_{opt}, at which the largest COD removals were achieved in treatment of industrial wastewaters of various strength conducted in the inverse fluidized bed biofilm reactor. The largest COD removals were achieved at the following air velocities u_{opt} and retention times t_s, and $(V_b/V_R) =$

0.55: 1) for COD_o = 72,780 mg/l at u_{opt} = 0.052 m/s and t_s = 80 h; 2) for COD_o = 62,070 mg/l at u_{opt} = 0.042 m/s and t_s = 65 h; 3) for COD_o = 49,130 mg/l at u_{opt} = 0.033 m/s and t_s = 55 h; 4) for COD_o = 41,170 mg/l at u_{opt} = 0.028 m/s and t_s = 45 h; 5) for COD_o = 35,460 mg/l at u_{opt} = 0.025 m/s and t_s = 27.5 h; and 6) for COD_o = 26,470 mg/l at u_{opt} = 0.014 m/s and t_s = 22.5 h. In the treatment operation conducted in a reactor optimally controlled at the above values of u_{opt}, t_s and (V_b/V_R), the following decreases in COD were obtained: 1) from 72,780 to 5410 mg/l; 2) from 62,070 to 3730 mg/l; 3) from 49,130 to 2820 mg/l; 4) from 41,170 to 1820 mg/l; 5) from 35,460 to 1600 mg/l; and 6) from 26,470 to 1180 mg/l, that is, approximately a 93%, 94%, 95%, 96%, 95% and 96% COD reduction was attained, respectively.

INTRODUCTION

Biological wastewater treatment conducted in systems based on activated sludge requires retention time of many days [1, 2]. On the other hand, the same treatment can be achieved in a fluidized bed biofilm reactor (FBBR) in retention time of several hours [1, 3]. The fluidized bed technology owes its high-rate success to much higher surface area and biomass concentration than those that can be achieved in the conventional treatment processes.

The FBBRs, in which the biomass is fixed on inert particles, are among the most effective apparatuses used in wastewater treatment [1]. The bed consisting of small particles offers a vast surface area for microbial growth in the state of fluidization. This enables far greater microbial concentration in a reactor than that maintained in the conventional fixed bed systems. The large biofilmliquid interfacial area, high interfacial velocities and good mass transfer characteristics are the main advantages of this type of reactors.

The three-phase (gas-liquid-solid) FBBR has been successfully applied in aerobic biological treatment of industrial and municipal wastewaters [4-8]. The reactor outperforms other reactor configurations used in wastewater treatment such as the activated sludge system and packed-bed (or trickling-filter) reactor [1,2]. The superior performance of the FBBR stems from the very high biomass concentration (up to 30 - 40 kg/m³) that can be achieved due to immobilization of cells onto or into the support media.

However, the excessive growth of biomass on the media can lead to washout of bioparticles (particles covered by biomass) from a reactor since the biomass loading can increase to such an extent that the bioparticles began to be carried over from the reactor [9,10]. The application of the inverse FBBR, in which a bed consisting of low density (matrix particle density smaller than that of liquid) particles expands downwards during fluidization, allows the control of biomass loading and provides the high oxygen concentration in the reacting liquid media [1,3].

In a FBBR containing low-density particles, fluidization can be conducted either by an upward co-current flow of gas and liquid through a bed (Figure 1) [3,10] or by a downward flow of liquid and countercurrent upward flow of gas [11,12]. In the former, fluidization is achieved by an upward flow of gas whereby the gas bubbles make the bed expanding downwards into the less dense mixture of gas and liquid. In the latter, the bed is fluidized by a downward flow of a liquid counter to the net buoyancy force of the particles. At a small flow of the liquid, not sufficient to counter to the net buoyancy force, fluidization can be achieved by an adequate upward flow of the gas. The fluidization where fluidized bed expands downwards is termed the inverse fluidization.

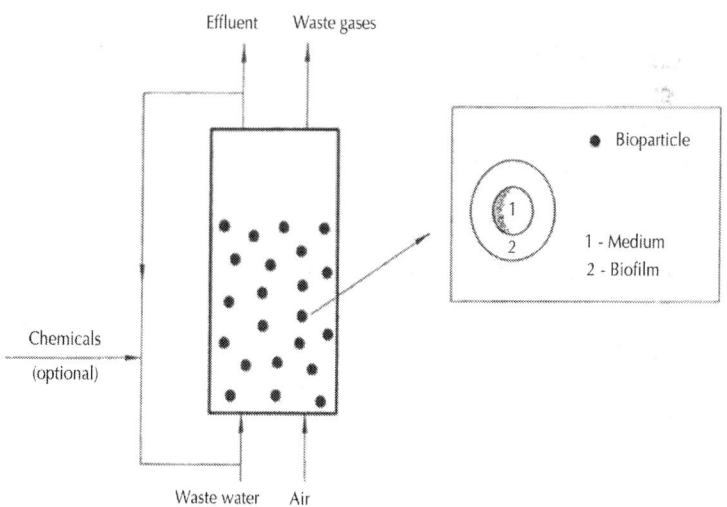

Figure 1: Scheme of the inverse fluidized bed biological reactor.

In this study, the optimal aeration in the inverse FBBR, for which the largest COD removals were achieved, was determined for treatment of industrial wastewaters of various strengths. The polypropylene particles of density 910 kg/m^3 were fluidized by an upward flow of air through a bed. Experiments on COD reduction were performed for the ratio of settled bed volume to reactor volume (V_b/V_R) = 0.55, and various air velocities u and residence times t. The ratio (V_b/V_R) = 0.55 was applied because for this value of (V_b/V_R) Sokół and Korpal [1], and Sokół et al. [3] have obtained the largest reduction in COD in treatment of refinery wastewaters.

EXPERIMENTATION

Experimental Set-Up

Experiments were conducted in the reactor shown in Figure 2. A growing medium, stored in a reservoir 1, was pumped into the bottom of the reactor by a centrifugal pump 5. Before entering the bed, the liquid was mixed with air by means of a sparger. The air was introduced to the bed through a distributor 7 whose plate had 200 holes of 4 mm diameter on a triangular pitch. The fluidised bed section 9, made of Duran glass, had a 20 cm internal diameter and was 6 m high. It was ended by a disengaging cap 10 with a 60 cm internal diameter and a height of 80 cm. The biomass sloughed off from the particles was separated from the effluent in a vessel 6 and removed from the system. The flow rate of the liquid was measured by a rotameter 4 and controlled by a ball valve. The air flow rate was measured using a rotameter 11 and controlled by a needle valve. The pH was adjusted by a control system 3, consisting of a pH-meter and micropumps supplying base or acid; as required. The temperature control system 2 consisted of a coil with cold water and an electric heater coupled with a contact thermometer.

The biomass support was the polypropylene particles of density 910 kg/m^3 which are described elsewhere [9].

Feed and Microorganisms

The growing medium was the wastewater (Feed 1) taken from a local refinery. The Feed 1 was diluted so as to obtain the Feeds 2-6 whose composition is given in Table 1. The wastewaters were enriched in mineral salts by adding the following (mg/l): $(NH_4)_2SO_4$—500; KH_2PO_4 —200; $MgCl_2$—30; NaCl—30; $CaCl_2$—20; and $FeCl_3$— 7 [1].

The inoculum was the activated sludge taken from the biological treatment unit operated at the refinery.

Methodology

Sokół and Korpal [1] have established that the optimal ratio (V_b/V_R) for a FBBR when used in biological wastewater treatment was equal to 0.55. Therefore, in this study experiments were performed for the ratio $(V_b/V_R) = 0.55$.

Biomass Culturing

The particles and the growing medium (Feed 6) were introduced into the reactor to give a ratio $(V_b/V_R) = 0.55$. To start growth of the microorganisms on the particles, a batch culture was first initiated by introducing about [5] of the inoculum into the reactor. Then the culture was incubated for approximately 48 h to encourage cell growth and the adhesion of freely suspended biomass on the particles. The air was supplied at the flow rate of 0.03 m^3/s and this was found to be sufficient for biomass growth [1, 3]. The pH was controlled in the range 6.5 - 7.0 and the temperature was maintained at 28˚C - 30˚C.

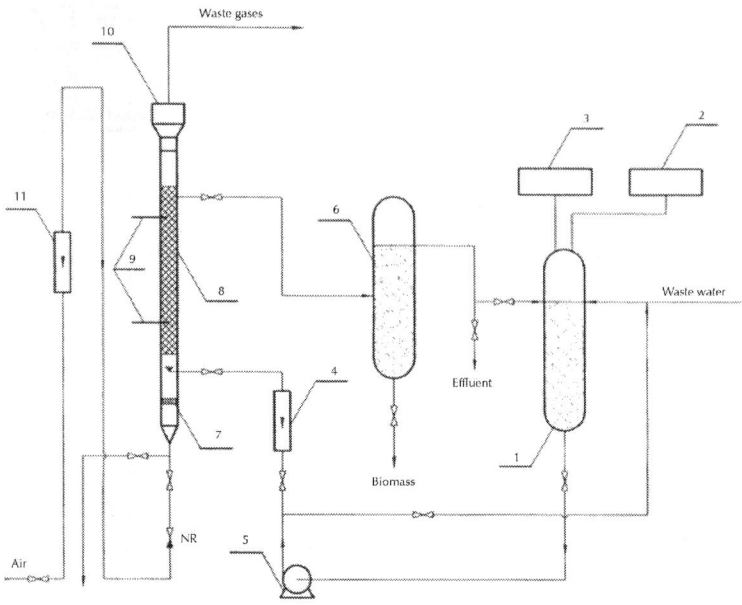

Figure 2: Schematic diagram of the experimental apparatus: 1. Reservoir; 2. Temperature control system; 3. pH control system; 4. Liquid rotameter; 5. Pump; 6. Intermediate reservoir; 7. Air distributor; 8. Sampling; 9. Fluidized section; 10. Disengaging section; 11. Air rotameter.

Table 1: Composition of wastewaters (feeds)

	Concentration ($\times 10^3$) mg·dm^{-3}					
Constituent	Feed 1	Feed 2	Feed 3	Feed 4	Feed 5	Feed 6
o-Cresol	16731	14286	11313	9492	7766	5951
m-Cresol	8944	7612	6028	5058	4180	3592
3,5 Dimethylphenol	7636	6496	5145	4316	3625	2659
Phenol	6343	5398	4275	3587	3058	2408

2,4-Dimethylphenol	5072	4446	3541	2954	2479	2012
Benzene	3774	3214	2545	2135	1755	1349
Toluene	3237	2757	2183	1832	1539	1278
Isopropylphenol	2523	2151	1703	1430	1220	947
3,4-Dimethylphenol	2415	2057	1629	1367	1114	878
o-Xylene	1158	986	781	655	538	472
2,6-Dimethylphenol	1132	964	763	641	525	441
C3-Phenyl	299	254	202	169	136	117
Ethylphenl	232	198	157	132	105	82
Cl-Phenyl	84	72	57	48	34	31

When the biofilm had begun to grow on the particles, the growing medium was started to be pumped into the reactor at a dilution rate D = 0.40 h^{-1}. This value of D corresponded to the smallest time t applied for Feed 6 (t = 1/D = 2.5 h in Figure 3). Next, the air velocity u was set at the smallest value applied for Feed 6 (u = 0.009 m/s in Figure 3) and the cultivation was continued until the constant biomass loading was achieved in a reactor.

The occurrence of the steady-state biomass loading was established by weighting the mass of cells grown on the support. The biomass was scraped from sample particles and dried at temperature 105°C for 30 minutes. It was considered that the steady state occurred when the weight of biomass in two consecutive samples differed less than 5%. The steady-state biomass loading was attained in a reactor after the cultivation for approximately two weeks.

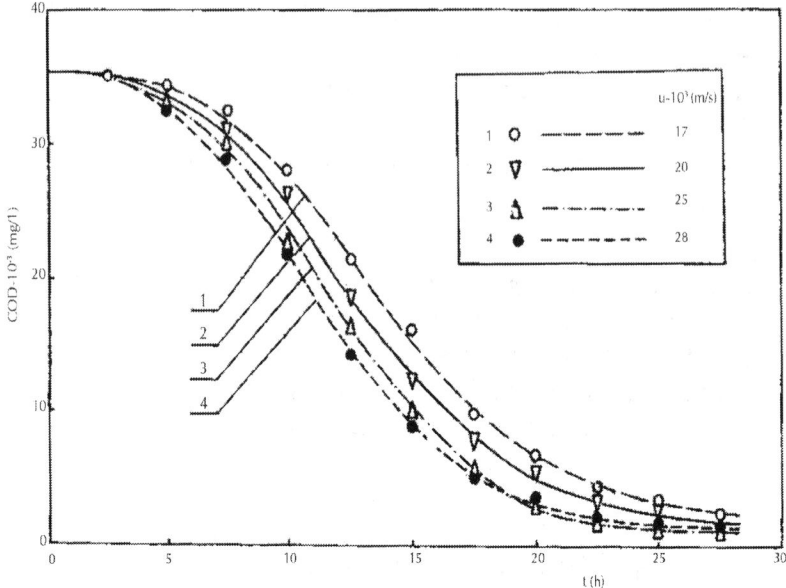

Figure 3: Dependence of chemical oxygen demand COD on residence time t for treatment of Feed 6.

Treatment Operation

When the steady-state biomass loading was achieved, a sample liquid was withdrawn from the reactor and COD was measured by the procedure recommended by Verstraete and van Vaerenbergh [13]. It was established that once the constant biomass loading occurred in a reactor, the value of COD was practically at steady state.

Next, the air velocity u was increased stepwise to its next value applied for Feed 6 (u = 0.011 m/s in Figure 3) and the cultivation was continued until the new steadystate biomass loading was achieved. When this was attained, COD was measured by the method mentioned earlier [13]. These experiments for Feed 6 were conducted for all values of u shown in Figure 3.

Then the dilution rate D was decreased stepwise to its next value applied for Feed 6 (t = 1/D = 5 h in Figure 3) and the air velocity u was re-set to its smallest value applied for Feed 6 (u = 0.009 m/s in Figure 3). The cultivation was continued until the steady-state biomass loading was achieved. When this occurred, COD was measured following the procedure mentioned above [13]. These experiments were conducted for all air velocities u and times t shown in Figure 3. The results are given in Figure 3.

The above experiments were also performed for Feeds 1-5 (Table 1). The results are shown in Figures 4-8.

It should be pointed out that the air velocities u applied in the experiments were several times larger than the minimum fluidization air velocity u_f. This was possible because the reactor was operated at the ratio $(V_b/V_R) = 0.55$ which was smaller than the critical ratio $(V_b/V_R)_{cr}$ [1, 9]. On the other hand, the air velocities u were smaller than the critical velocity u_{cr} at which the entire bed settled at the reactor bottom.

RESULTS AND DISCUSSION

It can be seen in Figures 3-8 that for a set time t, a concentration of COD in effluent depended on the air velocity u. A reduction in COD initially increased, and then decreased with an increase in u. For example, it can be noticed in Figure 4 that for a set t the values of COD were decreasing with an increase in u from 0.009 up to 0.014 m/s. A further increase in u did not improve a COD removal. The smallest value of COD was attained for $u_m = 0.014$ m/s. This can be explained by the fact that with an increase in u up to 0.014 m/s, an interfacial (air- liquid) area increased, and consequently the amount of the oxygen supplied for biomass growth increased [1,14]. Thus, for the u smaller than 0.014 m/s, oxygen was the limiting factor for biomass growth. On the other hand, for the air velocities greater than 0.014 m/s, the degradation rate of the constituents of the wastewaters was the controlling factor of the treatment process [3,15].

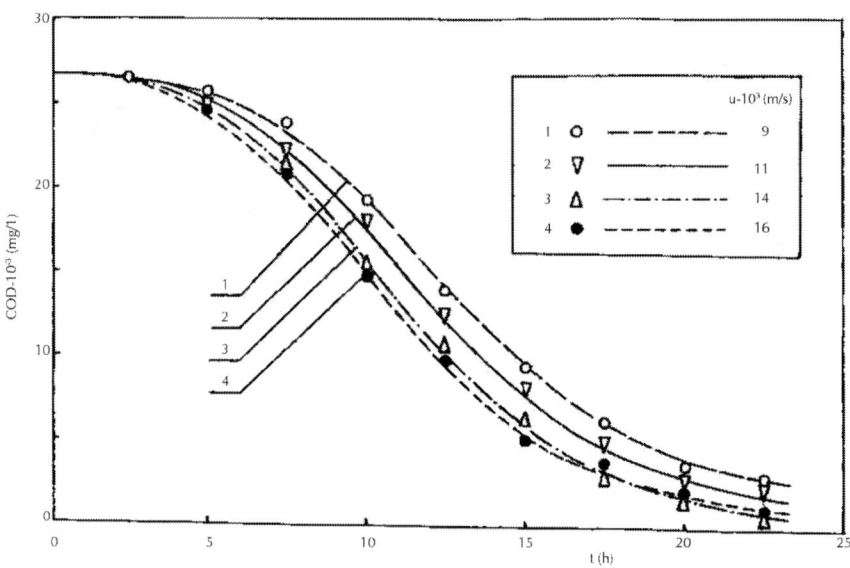

Figure 4: Dependence of chemical oxygen demand COD on residence time t for treatment of Feed 5.

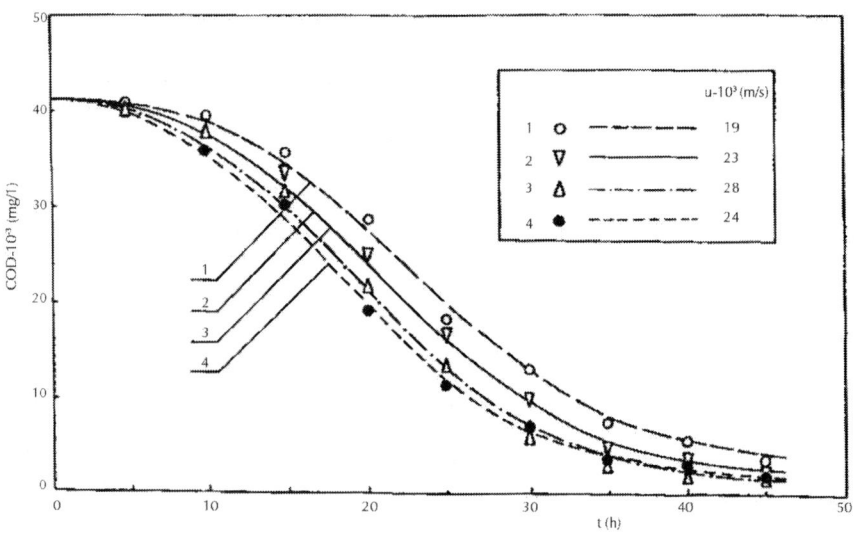

Figure 5: Dependence of chemical oxygen demand COD on residence time t for treatment of Feed 4.

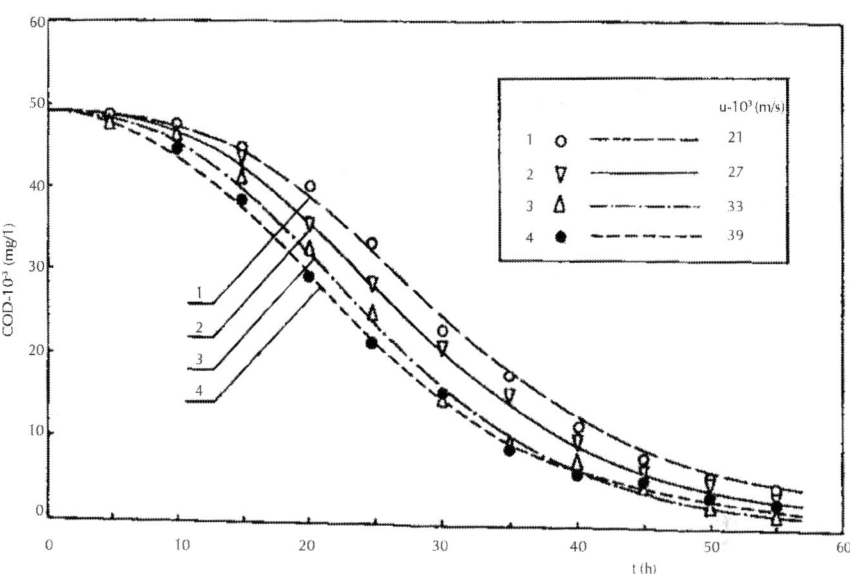

Figure 6 Dependence of chemical oxygen demand COD on residence time t for treatment of Feed 3.

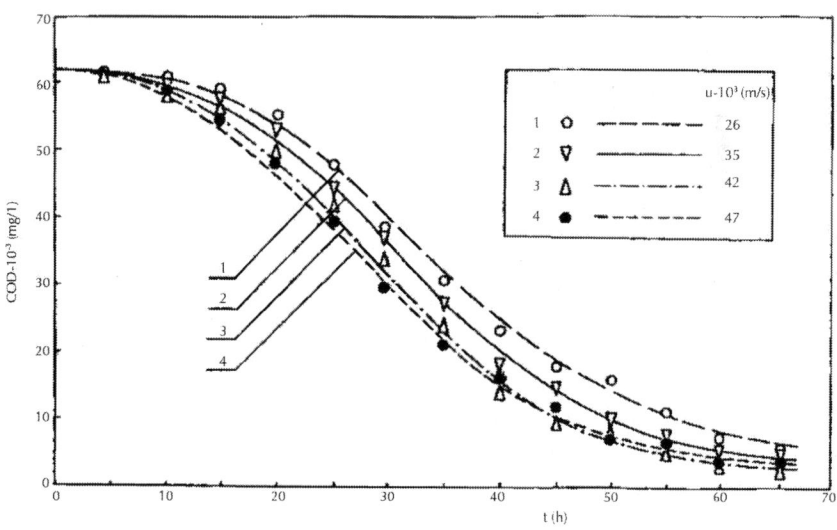

Figure 7: Dependence of chemical oxygen demand COD on residence time t for treatment of Feed 2.

As can be noticed in Figure 3, the value of COD was practically at steady state for times t greater than 22.5 h. The largest COD removal occurred when the reactor was operated at $(V_b/V_R) = 0.55$ and $u_{opt} = 0.024$ m/s. A decrease in COD from 26,470 to 1180 mg/l, that is, a 96% COD reduction, was achieved when a reactor was optimally controlled at $(V_b/V_R) = 0.55$, $u_{opt} = 0.014$ m/s and t = 22.5 h.

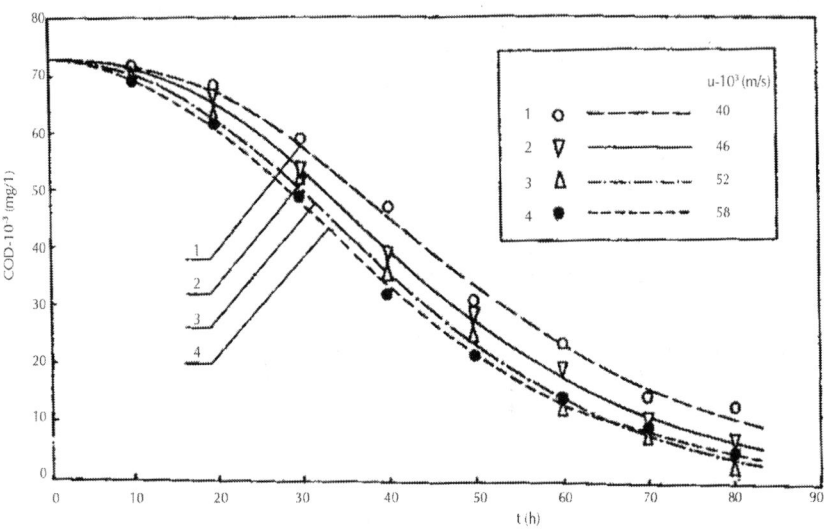

Figure 8: Dependence of chemical oxygen demand COD on residence time t for treatment of Feed 1.

The optimal air velocity u_{opt} for which the largest decrease in COD was obtained strongly depended on value of COD (Figures 4-8).

The biomass loading was successfully controlled in a reactor containing low density particles used as biomass support. This was due to particle geometry and particularly availability of the internal surface and the grooves on external surface of the particles for biomass growth. With such geometry of the particles, shear forces occurring between the particles and the liquid sloughed off excess of biomass mainly from the external, and to less extend from the internal, surface of the particles. Furthermore, attrition, associated with particle-particle and particle-wall collisions, of biomass grown in the grooves and on the internal surface was less abrupt than the cells grown on the external surface of the particles.

The steady-state biomass loading in a reactor depended on an air velocity u. After change in u, the new steady-state biomass loading occurred after the culturing for about 2 days.

CONCLUSIONS

The optimal operating parameters for a reactor when used in treatment of industrial wastewaters were as follows: u_{opt} = 0.052 m/s and t_s = 80 h for COD_o = 72,780 mg/l; u_{opt} = 0.042 m/s and t_s = 65 h for COD_o = 62,070 mg/l; u_{opt} = 0.033 m/s and t_s = 55 h for COD_o = 49,130 mg/l; u_{opt} = 0.028 m/s and t_s = 45 h for COD_o = 41,170 mg/l; u_{opt} = 0.025 m/s and t_s = 27.5 h for COD_o = 35,460 mg/l; and u_{opt} = 0.014 m/s and t_s = 22.5 h for COD_o = 26,470 mg/l.

In a reactor controlled at the optimal values of u_{opt} and t_s, the following COD reductions were obtained: from 72,780 to 5410 mg/l; from 62,070 to 3730 mg/l; from 49,130 to 2820 mg/l; from 41,170 to 1820 mg/l; from 35,460 to 1600 mg/l; and from 26,470 to 1180 mg/l, i.e. a 93%, 94%, 95%, 96%, 95% and 96% COD removal was attained, respectively.

REFERENCES

1. W. Sokół and W. Korpal, "Aerobic Treatment of Wastewaters in the Inverse Fluidised Bed Biofilm Reactor," Chemical Engineering Journal, Vol. 118, No. 3, 2006, pp. 199-205.doi:10.1016/j.cej.2005.11.013

2. P. Hüppe, H. Hoke and D. C. Hempel, "Biological Treatment of Effluents from a Coal Tar Refinery Using Immobilized Biomass," Chemical Engineering Technology, Vol. 13, No. 1, 1990, pp. 73-79. doi:10.1002/ceat.270130110

3. W. Sokół, A. Ambaw and B. Woldeyes, "Biological Wastewater Treatment in the Inverse Fluidised Bed Reactor," Chemical Engineering Journal, Vol. 150, No. 1, 2009, pp. 63-68. doi:10.1016/j.cej.2008.12.021

4. A. Lohi, M. Aivarez-Cuenca, G. Anania, S. R. Upreti and L. Wan, "Biodegradation of Diesel Fuel-Contaminated Wastewater Using a Three-Phase Fluidized Bed Reactor," Journal of Hazardous

Materials, Vol. 154, No. 1-3, 2008, pp. 105-111.doi:10.1016/j. jhazmat.2007.10.001

5.   M. Bajaj, C. Gallert and J. Winter, "Biodegradation of High Phenol Containing Synthetic Wastewater by an Aerobic Fixed Bed Reactor," Bioresource Technology, Vol. 99, No. 17, 2008, pp. 8376-8381. doi:10.1016/j.biortech.2008.02.057

6.   P. A. Fitzgerald, "Comprehensive Monitoring of a Fluidized Bed Reactor for Anaerobic Treatment of High Strength Wastewater," Chemical Engineering Science, Vol. 51, No. 11, 1996, pp. 2829-2834. doi:10.1016/0009-2509(96)00160-1

7.   R. Sowmeyan and G. Swaminathan, "Evaluation of Inverse Anaerobic Fluidized Bed Reactor for Treating High Strength Organic Wastewater," Bioresource Technology, Vol. 99, No. 9, 2008, pp. 3877-3880. doi:10.1016/j.biortech.2007.08.021

8.   R. Sowmeyan and G. Swaminathan, "Performance of Inverse Anaerobic Fluidized Bed Reactor for Treating High Strength Organic Wastewater During Start-Up Phase," Bioresource Technology, Vol. 99, No. 14, 2008, pp. 6280-6284.doi:10.1016/j. biortech.2007.12.001

9.   W. Sokół and M. R. Halfani, "Hydrodynamics of a GasLiquid-Solid Fluidized Bed Bioreactor with a Low Density Biomass Support," Biochemical Engineering Journal, Vol. 3, No. 3, 1999, pp. 185-192. doi:10.1016/S1369-703X(99)00016-9

10.   W. Sokół, "Operational Range for a Gas-Liquid-Solid Fluidized Bed Aerobic Biofilm Reactor with a Low-Density Biomass Support," International Journal of Chemical Reaction Engineering, Vol. 8, 2010, Article A111. http://www.bepress.com/ijcre/vol8/ A111.

11.   L. Nikolov and D. Karamanev, "Experimental Study of the Inverse Fluidised Bed Biofilm Reactor," Canadian Journal of Chemical Engineering, Vol. 65, 1987, No. 2, pp. 214-217.doi:10.1002/ cjce.5450650204

12.   N. Fernandez, S. Montalvo, R. Borja, L. Guerrero, E. Sanchez, I. Cortes, M. F. Comenarejo, L. Traviso and F. Raposo, "Performance Evaluation of an Anaerobic Fluidized Bed Reactor with Natural Zeolite as Support Material When Treating High-Strength Distillery Wastewater," Renewable Energy, Vol. 33, No. 11, 2008, pp. 2458-2466.doi:10.1016/j.renene.2008.02.002

13. W. Verstraete and E. van Vaerenbergh, "Aerobic Activated Sludge," In: W. Schonborn, Ed., Biotechnology, VCH Verlagessellschaft GmbH, Weinheim, 1986, pp. 43- 112.

14. D. G. Karamanev, T. Nagamune and K. Endo, "Hydrodynamics and Mass Transfer Study of a Gas-Liquid-Solid Draft Tube Spouted Bed Bioreactor," Chemical Engineering Science, Vol. 47, No. 13-14, 1992, pp. 3581-3588. doi:10.1016/0009-2509(92)85073-K

15. W. Sokół and B. Woldeyes, "Evaluation of the Inverse Fluidized Bed Biological Reactor for Treating HighStrength Industrial Wastewaters," Advances in Chemical Engineering and Science, Vol. 1, No. 4, 2011, pp. 239- 244.

Chapter 6

Heavy Metals Accumulation in Soil Irrigated with Industrial Effluents of Gadoon Industrial Estate, Pakistan and its Comparison with Fresh Water Irrigated Soil

Noor Amin[1], Dawood Ibrar[1], and Sultan Alam[2]

[1]Department of Chemistry, Abdul Wali Khan University, Mardan, Pakistan

[2]Department of Chemistry, University of Malakand, Chakdara, Pakistan

ABSTRACT

Wastewater mixed with industrial effluents is used for irrigation in Gadoon Industrial estate and thus contaminating soil. This soil was tested for heavy metal content by using atomic absorption spectrophotometer (perkin elmer 700) and compared with control soil irrigated with tube well water at seven selected spots. Accumulation of the toxic metal was significantly greater in the soil irrigated with industrial effluent than control soil ($p < 0.05$). Manganese (Mn) was the most significant pollutant, accumulated up to 9.95 ppm in the soil irrigated with industrial waste water. It was found that the samples were containing Zn in the range of 1.596 - 6.288, Cu 0.202 - 1.236, Co 0.074 - 0.115, Ni 0.0002 - 0.544, Cr 0.243 - 0.936, Mn 3.667 - 9.955 and Pb 0.488 - 1.259 ppm. No sample was containing the heavy metal above the critical level mentioned in typical and unsafe heavy metal levels in soil.

INTRODUCTION

Industry is the backbone of a country for its development and with the growing population, the need for establishing new industries is increasing. Industries on one side manufacture useful products but on the other side time generate waste products, causing various environmental problems. The waste products may be in the form of solid, liquid or gas which leads to the creation of hazards, pollution and losses of energy. The wastes containing different pollutants and heavy metals are discharged into water and soil and ultimately pose a serious threat to human and ecosystem. By-products of different industries like textile, metal, dying chemicals, fertilizers, pesticides, cement, petrochemical, energy and power, leather, sugar processing, construction, steel, engineering and food processing industries are the main contributors to the soil pollution. So the rapid industrialization is ac- companied by both direct and indirect adverse effects on environment [1]. Industrial development may result in the generation of industrial effluents. It has been studied that many industries discharge untreated effluents into river and only 10% industries surveyed had primary treatment ranging from oxidation tanks, sedimentation tanks in developing countries [2] [3] .

Heavy metal toxicity may results into a number of health problems including a damaged or reduced mental and central nervous function, lower energy levels, and damage to blood composition, lungs, kidneys, liver, and other vital organs. Long-term exposure may result in a slowly progressing physical, muscular, and neurological degenerative process, that mimic Alzheimer's disease, Parkinson's disease, muscular dystrophy, and multiple sclerosis. Allergies are not uncommon and repeated long-term contact with some metals or their compounds may even cause cancer [4] [5].

Heavy metals on the basis of their health importance can be classified into four major groups, as essential, like Cu, Zn, CO, Cr, Mn and Fe, which are micronutrients and are toxic when taken in excess [6] [7] , non-essential like Ba, Al, Li and Zr, less toxic like Sn and Al, and highly toxic like Hg and Cd. The toxicity limit and recommended or safe intake of some of heavy metals for human health is given in Table 1.

In small quantities, certain heavy metals are nutritionally essential for a healthy life. Some of these are referred to as the trace elements (e.g., iron, copper, manganese, and zinc). These elements, or some form of them, are commonly found naturally in foodstuffs, in fruits and vegetables, and in commercially available multivitamin products [8] - [10] .

Because of the rapid industrialization, soil pollution by heavy metals is becoming a serious problem. Being an ultimate sink for industrial wastes, almost all industrial wastes are dumped into soil. Heavy metals in wastes find specific adsorption sites in soil where they are retained relatively stronger either on inorganic or organic colloids. [11] - [13]. Research has proven that long term use of sewage effluent for irrigation contaminates soil and crops to such an extent that it becomes toxic to plants and causes deterioration of soil. This contains considerable amount of potentially harmful substances including soluble salts like Fe^{2+}, Cu^{2+}, Zn^{2+}, Mn^{2+}, Ni^{2+} and Pb^{2+}. Additions of these heavy metals are Undesirable [14] [15].

The properties of the soil along with the climate change also changes due to anthropogenic impact. The influence of acid rains on soils and sorption properties of soil has been extensively studied by scientists from various disciplines. In almost all cases, they found that acid rains decrease the ability of binding heavy metals to soil particles. However, for naturally high acidic soils or very weak soils like rusty soils, the effect of acid rains on soils is shown to be much smaller [15].

The present study was conducted with an aim to study the impact of industrial effluents of different industries of Gadoon industrial estate

Pakistan, carrying different heavy metals in them and absorbed in soil during irrigation. In this study only seven elements like Cu, Zn, Co, Cr, Pb, Ni and Mn were studied. The results of analysis collected from different locations have been compared and reported.

EXPERIMENTAL

Soil samples were collected from seven different sites of the Gadoon industrial estate Gadoon amazai swabi.

Table 1: Toxic limit/recommended/safe intake of heavy metals [12]

Heavy metal	Toxic limit	Recommended intake/Safe intake
Arsenic	3 mg/day for 2 - 3 weeks	15 - 25 micro g/day (adults)
Cadmium	200 micro/kg of fresh weight	15 - 50 micro g/day (adults), 2 - 25 micro g/day children
Lead	>500 micro g/L (Blood)	20 - 280 micro g/day (adults), 10 - 275 micro g/ day children
Zinc	150 micro/day	15 micro g/day

The samples identification along with their location is shown in Table 2. From each sampling site five soil samples were collected from different location and mixed together. The samples were dried in the sun for four days, grinded and sieved. The samples were then reduced to laboratory samples by using the tabling process. The samples were named as S_1, S_2, S_3, S_4, S_5, S_6 and S_7. The samples were then dried in oven at 110°C for about five hours to remove moisture completely. The chemicals used were Nitric acid, Per-chloric acid, distilled water.

Nitric-Perchloric Acid Digestion

Nitric-perchoric acid digestion method was performed for sample preparation [7]. One gram of a sample was placed in 250 ml digestion tube and 10 ml of concentrated HNO_3 was added. The mixture was boiled for 30 - 45 minutes to oxidize all easily oxidizable matter. After cooling, 5 ml of 70% $HClO_4$ was added and the mixture was boiled gently till the appearance of dense white fumes. The contents were cooled and 20 ml of distilled water was added, and re boiled to stop the release of any fumes. The solution was cooled again, filtered off through Whatman No. 42 filter paper and transferred to 25 ml volumetric flask. The volume was made up to the mark with distilled water. Blank solution was prepared with the same procedure except the addition of soil sample.

Standards for different elements were prepared from the stock solutions (1000 ppm) using dilution method. For each element different dilutions were made for calibration curve given in Table 3.

Analysis of Trace Metals

Trace metals such as Cu, Co, Fe, Pb, Cr, Mn, Zn and Ni were analyzed in all the samples using atomic absorption spectrophotometer.

Statistical Analysis

Data obtained during current study was analyzed statistically for mean, standard deviation, ANOVA and Dun- can Multiple Range Test (DMRT) by using SPSS for windows, version 16.0 (SPSS Inc., Chicago, IL, USA). Probability less than 0.05 was accepted as significant.

Table 2: Locations at Gadon Amazai from where different samples were collected

S. No	Sample ID	Sample Location
1	S1	2 km from Sardar Chemical Industries
2	S2	2500 m Sardar Chemical Industries and Cherat Paper Sack
3	S3	1 km Shafi Chemical Industries and Hamza Steel Industries
4	S4	500 m Shafi Chemical Industries Plot-2
5	S5	500 m T.W Metal Recycling Industries and Poyal Jadoon Marble Factory

| 6 | S6 | 1 km T.W Metal Recycling Industries Plot 11/16 |
| 7 | S7 | 200 m Shafi Enterprises |

Table 3: Standards used for different metals

S/No	Standard Name	Standard Symbol	Concentration (ppm)		
1	Copper	Cu	2	4	8
2	Zinc	Zn	1	2	4
3	Cobalt	Co	7	14	21
4	Nickel	Ni	7	14	28
5	Manganese	Mn	2.5	5	10
6	Chromium	Cr	2	4	8
7	Lead	Pb	12.5	25	50

RESULTS AND DISCUSSION

The concentration of copper in soil of the study area ranged from 0.202 to 1.236 ppm (Figure 1). The highest amount of this metal was present at location 5, situated on a distance of 500 m from T.W Metal recycling Industries and Poyal Jadoon Marble Factory (Table 2). Accumulation of this metal in soil irrigated with industrial effluent was significantly greater than its amount in soil irrigated with tube well water ($p < 0.05$). The Typical background levels for non-contaminated soil is 1 - 50 ppm, if the concentration of copper in the soil exceeds 200 ppm it become unsafe for leafy vegetable while the concentration of copper greater than 500ppm become unsafe for garden and children contact [14] . The results show that the study area contains the copper metal in a permissible range.

The concentration of zinc in the studied area varied from 1.596 to 6.288 ppm and its average concentration was 2867 ppm (Figure 2). At all locations, the accumulation of Zn was significantly greater in soil irrigated with wastewater than control coil ($p < 0.05$). Zinc is the second most abundant heavy metal found in the study area. The typical background levels of zinc for non-contaminated soil is very broad, i.e. 9 - 125 ppm, while its concentration above 200 ppm is unsafe for leafy or root vegetables. If the concentration of zinc in soil exceeds 500 ppm, it is considered unsafe for gardens and children contact [15]. The results of the present study show that the study area contains zinc in permissible range.

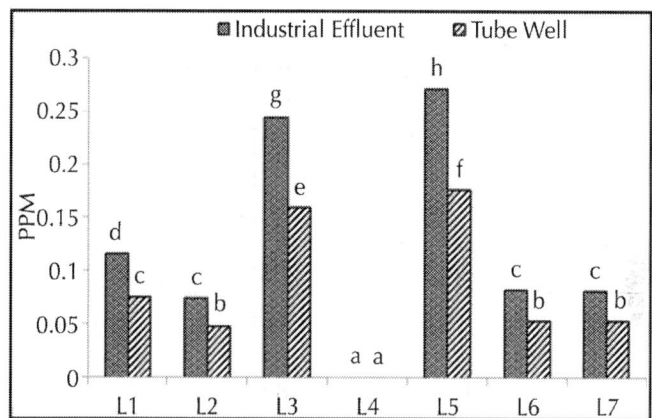

Figure 1: Accumulation of cobalt in soil irrigated with industrial effluent and tube well water at different locations in Gadoon Amazai industrial zone, Pakistan. Bars represents mean value of three replicates and bars labeled with different letters are significantly different from each other (Duncan Multiple Range test; $p < 0.05$).

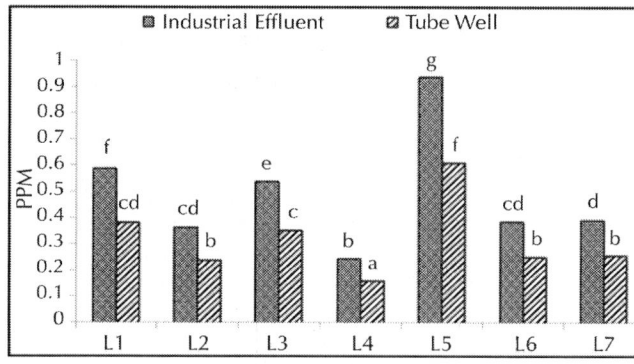

Figure 2: Accumulation of chromium in soil irrigated with industrial effluent and tube well water at different locations in Gadoon Amazai industrial zone,

Pakistan. Bars represents mean value of three replicates and bars labeled with different letters are significantly different from each other (Duncan Multiple Range test; $p < 0.05$).

The concentration of cobalt was in the range of 0.001 to 0.271 ppm at different locations in the study area and its average concentration was 0.124 ppm (Figure 3). The highest concentration of this metal was also in the soil irrigated with contaminated water at location 5 ($p < 0.05$). Concentration of this metal was well below the toxic level in the study area and minimum among all metals detected over there.

From Figure 4, it is obvious, that the concentration of Nickel in the studied area was below the detection limit in soil from L1 and L2. Amount of this metal was greatest (0.38 - 0.544 ppm) in wastewater irrigated soil of location 5 and 6, which was significantly greater than the concentration of Ni in other locations ($p < 0.05$). Although the abundance of Ni was significantly greater in soil supplied with wastewater than soil irrigated with tube well water (Figure 5), nevertheless, it was well below the critical levels of this metal [15]. Nickle is the sixth abundant heavy metal found in the study area. The typical background levels of nickel for non-contaminated soil is very broad, i.e. 0.5 - 50 ppm, while its concentration above 200 ppm is unsafe for leafy or root vegetables. If the concentration of nickel in soil exceeds 500 ppm, it is considered unsafe for gardens and children contact [15] [16].

Chromium was the fifth abundant heavy metal in all the locations of the study area (Figure 6). Its concentration varied from 0.243 in sample-4 to 0.936 ppm in sample-5 and its average concentration was found to be 0.493 ppm. Soil of location 5 had the greatest concentration of this metal ($p < 0.05$), irrespective of the water source used for irrigation.

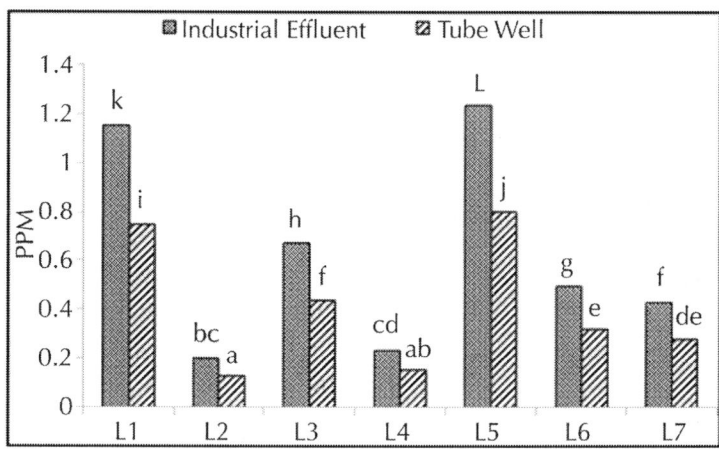

Figure 3: Accumulation of copper in soil irrigated with industrial effluent and tube well water at different locations in Ga- doon Amazai industrial zone, Pakistan. Bars represents mean value of three replicates and bars labeled with different letters are significantly different from each other (Duncan Multiple Range test; p < 0.05).

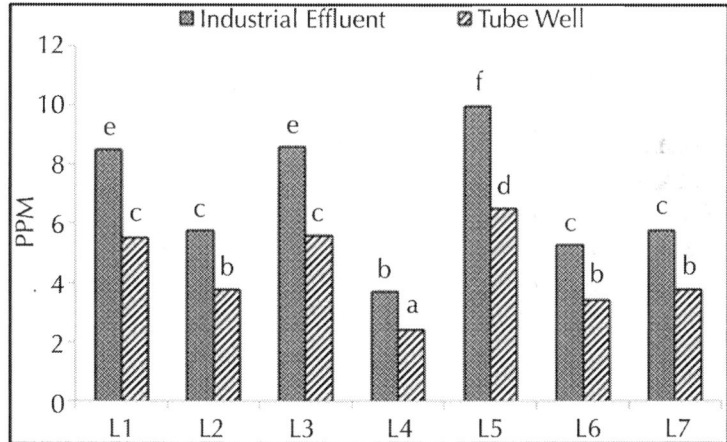

Figure 4: Accumulation of manganese in soil irrigated with industrial effluent and tube well water at different locations in Gadoon Amazai industrial zone, Pakistan. Bars represents mean value of three replicates and bars labeled with different letters are significantly different from each other (Duncan Multiple Range test; p < 0.05).

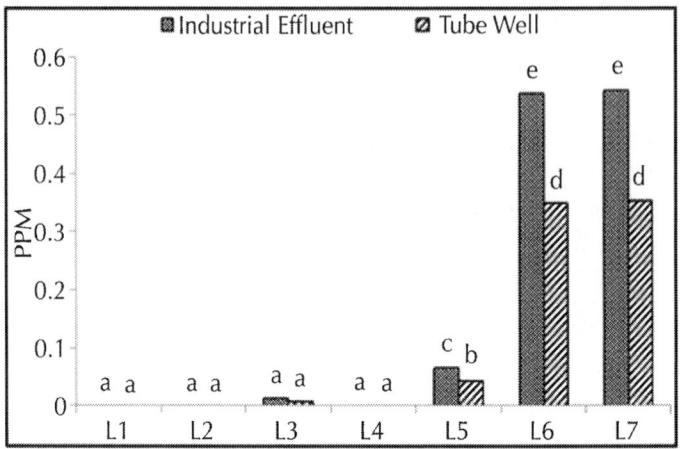

Figure 5: Accumulation of nickel in soil irrigated with industrial effluent and tube well water at different locations in Ga- doon Amazai industrial zone, Pakistan. Bars represents mean value of three replicates and bars labeled with different letters are significantly different from each other (Duncan Multiple Range test; $p < 0.05$).

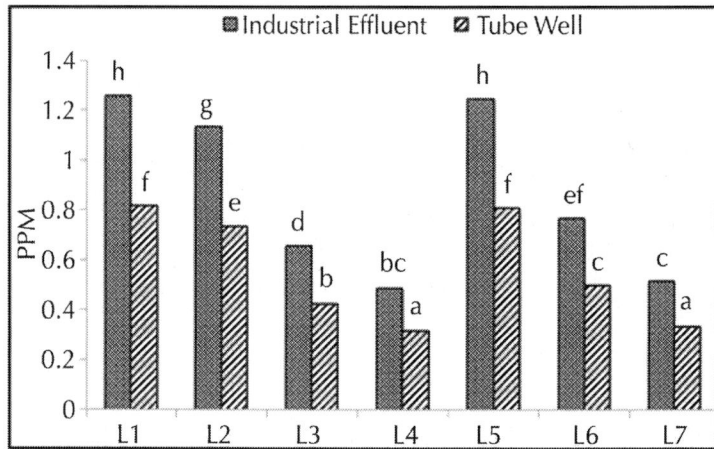

Figure 6: Accumulation of lead in soil irrigated with industrial effluent and tube well water at different locations in Gadoon Amazai industrial zone, Pakistan. Bars represents mean value of three replicates and bars labeled with different letters are significantly different from each other (Duncan Multiple Range test; $p < 0.05$).

The most abundant heavy metal found in the study area was manganese which varied from 3.667 in sample-4 to 9.955 ppm in sample-5 (Figure 7).

Lead in the study area varied from 0.488 in sample 4 to 1.259 ppm in sample (Figure 8). Lead was the third most abundant heavy metal found in our study area. The Typical background level of lead for non-contaminated soil is 10 - 70 ppm. When the concentration of lead goes above 500 ppm, it become unsafe for leafy or root vegetables. If the concentration of zinc in soil exceeds 1000 ppm, it is then considered unsafe for gardens and children contact [17] [18]. The result of the present study shows that the study area contains zinc in permissible range. Overall the concentration of heavy elements in the study area are given in the order Mn > Zn > Pb > Cu > Cr > Ni > Co.

The result showed a high level of zinc and manganese and low level of cobalt and nickel in soil samples. The variations of the metal contents observed in these soil samples depends on the physical and chemical nature of the soil and absorption capacity of each metal in the soil which is altered by the innumerable environmental factors and nature of soil (Reference). Location 5 contained greater amount of heavy metals, which was elevated to even greatest by irrigation with industrial effluent. Although at all locations, the metals were well below their critical values; however, continuous monitoring of the soil may help to predict any increase in future.

Soils normally contain low background levels of heavy metals. Excessive levels of heavy metals can be hazardous to man, animals and plants. Heavy metals regulated by the EPA are arsenic, cadmium, copper, lead, nickel, selenium, and Zinc. The limits of some heavy metals for non-contaminated soil and plants and garden are given in Table 4. Unsafe levels of heavy metals affect the soil texture, organic matter and pH.

CONCLUSIONS

The present study showed that heavy metals found in soil irrigated by the effluents of Gadoon industrial area are in the order of Mn > Zn > Pb > Cu > Cr > Ni > Co and their concentration is below the typical background level for non-contaminated soil as given in Table 4. This shows that plants and vegetables grown in this soil should have no adverse effects on animals and human beings. However, it may be emphasized that prolong use of industrial effluent for irrigation may lead to accumulation of heavy metals to toxic level in the soil.

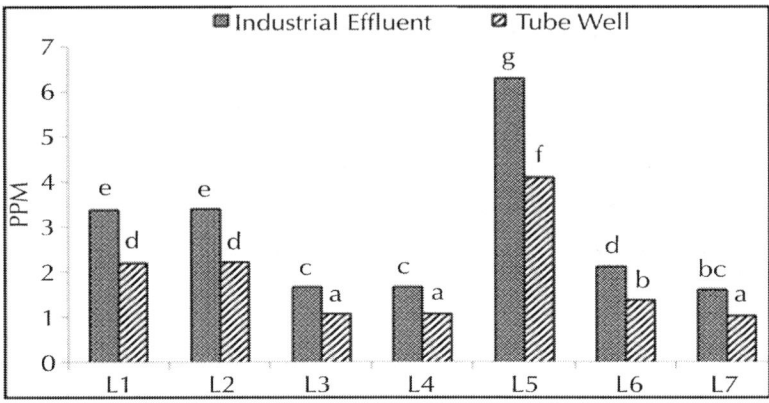

Figure 7: Accumulation of zinc in soil irrigated with industrial effluent and tube well water at different locations in Gadoon Amazai industrial zone, Pakistan. Bars represents mean value of three replicates and bars labeled with different letters are significantly different from each other (Duncan Multiple Range test; $p < 0.05$).

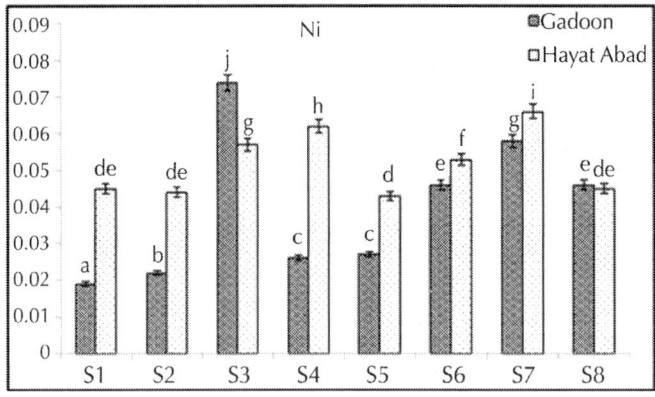

Figure 8: Mean concentration of Ni in the discharge of various industries found in the industrial area of Gadoon Amazai, Sa- wabi and Hayat Abad, Peshawar. Bars labeled with different alphabets are significantly different from each other (Duncan Multiple Range Test; $p < 0.05$).

Table 4: Typical and unsafe heavy metal soil levels [13]

Heavy metal	Typical background levels for non-contaminated soil (ppm)	Unsafe for leafy or root vegetables (ppm)	Unsafe for gardens and children contact (ppm)
Cadmium	0.1 - 1.0	>10	50
Copper	1 - 50	>200	500
Lead	10 - 70	>500	1000
Nickel	0.5 - 50	>200	500
Zinc	9 - 125	>200	500

REFERENCES

1.   Amin, N., Hussain, S., Alamzeb, A. and Begum, S. (2013) Accumulation of Heavy Metals in Edible Parts of Vegetables Irrigated with Waste Water and Their Daily Intake to Adults and Children, District Mardan, Pakistan. Food Chemistry, 136, 1515-1532.http://dx.doi.org/10.1016/j.foodchem.2012.09.058

2.   Amin, N., Shah, M.T. and Ali, K. (2009) Investigation of Raw Material for the Manufacturing of Sulphate-Resisting Cement in Darukhula Nizampur, NWFP, Pakistan. Magazine of Concrete Research, 61, 779-785.http://dx.doi.org/10.1680/macr.2008.61.10.779

3.   Ali, K., Amin, N. and Shah, M.T. (2008) Chemical Study of Limestone and Clay for Cement Manufacturing in Da- rukhula, Nizampur District, Nowshera, North West Frontier Province (N.W.F.P.), Pakistan. Chinese Journal of Geo- chemistry, 27, 242-248.http://dx.doi.org/10.1007/s11631-008-0242-8

4.   Wang, L.X., Guo, Z.H., Xiao, X.Y., Chen, T.B., Liao, X.Y. and Song, J. (2008) Variations and Trends of the Freezing and Thawing Index along the Qinghai-Xizang Railway for 1966-2004. Journal of Geographical Sciences, 18, 3-16.http://dx.doi.org/10.1007/s11442-008-0003-y

5.   Rattan, R.K., Datta, S.P., Chhonkar, P.K., Suribabu, K. and Singh. A.K. (2005) Long-Term Impact of Irrigation with Sewage Effluents on Heavy Metal Content in Soils, Crops and Groundwater—A Case Study. Agriculture, Ecosystems & Environment, 109, 310-322.http://dx.doi.org/10.1016/j.agee.2005.02.025

6.   Amin, N. (2010) Chemical Activation of Bagasse Ash in Cementitious System and Its Impact on Strength Development. Journal of the Chemical Society of Pakistan, 32, 481-484.

7.   Sharma, R.K., Agarwal, M. and Marshall, F. (2007) Heavy Metals Contamination of Soil and Vegetables in Suburban Areas of Varanasi, India. Ecotoxicology and Environmental Safety, 66, 258-266.

8.   Sridhara, N., Chary, C.T. and Samuel, S.R.D. (2008) Asian Journal of Ecotoxicology, 69, 43.

9.   Zhou, S.L., Lu, C.F. and Wan, H.Y. (2005) Journal of Henan Normal University (Natural Science), 33, 12.

10.   Xie, Z.M., Li, J., Chen, J.J. and Wu, W.H. (2006) Asian Journal of Ecotoxicology, 1, 2.

11.   Zhang, C.S. (2006) Using Multivariate Analyses and GIS to Identify Pollutants and Their Spatial Patterns in Urban Soils in Galway, Ireland. Environmental Pollution, 142, 501-511.http://dx.doi.org/10.1016/j.envpol.2005.10.028

12.   US EPA National Recommended Water Quality Criteria. Federal Register, Care (1998).

13.   Cui, Y.J., Zhu, Y.G., Zhai, R.H., Chen, D.Y., Huang, Y.Z. and Qiu, Y. (2004) Environment International, 3, 25.

14.   Ding, A.F., Pan, G.X. and Li, L.Q. (2006) Distribution of PAHs in Particle-Size Fractions of Selected Paddy Soils from Tai Lake Region, China and Its Environmental Significance. Acta Sci. Circum., 26, 293-299.

15.   Melloul, O.L., Hassani, A. and Bouhoum, K. (2002) International Journal of Environmental Health, 23, 21.

16.   Fleck, J.A., Grigal, D.F. and Nater, E.A. (1999) Mercury Uptake by Trees: An Observational Experiment. Water Air and Soil Pollution, 115, 513-523.http://dx.doi.org/10.1023/A:1005194608598

17.   Khan, S., Cao, Q., Zheng, Y.M., Huang, Y.Z. and Zhu, Y.G. (2008) Health Risks of Heavy Metals in Contaminated Soils and Food

Crops Irrigated with Wastewater in Beijing, China. Environmental Pollution, 152, 686-692. http://dx.doi.org/10.1016/j.envpol.2007.06.056

18. Lupton, G.P., Kao, G.F., Johnson, F.B., Graham, J.H. and Helwig, E.B. (1985) Cutaneous Mercury Granuloma. Jour- nal of American Academy of Dermatology, 12, 296-303.http://dx.doi.org/10.1016/S0190-9622(85)80039-6

Optical-Electronic Properties of Carbon-Nanotubes Based Transparent Conducting Films

Kuan-Ru Chen[1], Hsiu-Feng Yeh[1], Hung-Chih Chen[1], Ta-Jo Liu[1*], Shu-Jiuan Huang[2], Ping-Yao Wu[2], and Carlos Tiu[3]

[1]Department of Chemical Engineering, National Tsing Hua University, Hsinchu, Chinese Taipei

[2]Material and Chemical Laboratory, Industrial Technology & Research Institute, Hsinchu, Chinese Taipei

[3]Department of Chemical Engineering, Monash University, HHMelbourneHH, Australia

ABSTRACT

Three coating methods (slot, dip and blade coatings) were used separately to coat a well-dispersed single-wall carbonnanotube (SWCNT) solution on polyethylene terephthalate (PET) film, and the resulting optical and electronic properties were measured and compared. It was found that the sheet resistance and the transparency of the SWCNT coated film decreased as the coating speed increased for dip and blade coatings, but were independent of the coating speed for slot coating. All three coating methods were able to produce transparent conductive film with transparency above 85% and sheet resistance close to 1000 ohm/sq. For industrial production, the slot die coating method appears to be more suitable in terms of high coating speed and uniformity of optical and electronic properties.

INTRODUCTION

Transparent conductive film or glass is a key component for many optical-electronic devices such as organic light emitting diodes (OLED), organic photovoltaic solar cells (OPV), liquid crystal display (LCD) panels and touch panels, just to name a few. The most critical requirements for transparent conductive film or glass are low sheet resistance and high transparency. Many materials were considered suitable for making transparent conductive film or glass [1-4]. Owing to manufacturing and quality requirements, only indium tin oxide (ITO) film or glass has been commonly used for optical-electronic devices in the current market [5]. Despite its popularity, indium is a rare-earth material, and its price is high [6]. Furthermore, ITO films or glass has to be produced with a vacuum deposition technology. This technology is relatively expensive compared to conventional wet coating processes. Hence, many competitive approaches have been sought to replace ITO film or glass using different materials and coating methods.

Low-cost wet coating processes have been found to be promising in coating nano-scaled conductive media such as silver nanowires [7] or carbon-nanotubes [8-11] on PET film. Silver nanowires can be produced through an efficient chemical approach [12]. However, issues such as how to disperse silver nanowires and reduce their diameter for lower resistance still have to be resolved. Conductive film or glass coated with carbon-nanotubes (CNT) has similar optical-electronic properties, but is also hindered by the dispersion problems. Several effectiveapproaches have been attempted to overcome this issue for multi-walled and single-wall CNT [13-15], but the single-wall CNT (SWCNT) appeared to give better performance [16].

Spin coating is usually applied as a convenient means to prepare samples for laboratory analysis. However, usually over 90% of the coating solution is wasted, and is therefore not suitable for mass production. Recently, several researchers considered different coating methods for CNT solutions. Kim et al. [17] applied spin and spray coating methods for CNT electrode to make organic solar cells. de Andrade et al. [18] compared different technologies for the preparation of CNT networks. They concluded that dip coating and electrophoretic deposition are promising methods for solar cell application. In the present study, three different coating methods were used to make transparent conductive film with a well-dispersed single-wall CNT solution, and the optical and electronic properties of the samples were measured and compared.

The optical requirement for conductive films is that it must be over 85% transparent. The sheet resistance may however vary, depending on special applications. In the present study, it is chosen to be 1000 ohm/sq, which meets the requirement of electrostatic dissipation. The results presented here would be useful for future mass production considerations.

EXPERIMENTAL

Preparation of SWCNT Solution

Single walled carbon nanotubes (SWCNT) were prepared by the floating method in a vertical tube reactor [19] by using alcohol as the carbon source. The alcohol solution with a given composition of ferrocene and thiophene was introduced into the reactor with hydrogen as the carrier gas. The typical reactor temperature was between 1000°C - 1200°C. The SWCNT produced were purified by combining two-step processes of thermal annealing in air and acid treatment [20]. The SWCNT produced had average diameters around 2 to 2.5 nm, with purity >90% based on TGA and G/D ratio (Raman) around 35. The aqueous SWCNT dispersion was prepared by ultrasonication using a tip sonicator with sodium dodecyl benzene sulfonate (SDBS) as surfactant. The concentrations of the SWCNT was 0.1% and the ratio of SWCNT to SDBS was 1:1.5.

Coating Methods

Three coating devices were selected for making samples. The first device was a laboratory blade coater (Zehntner, ZUA 2000), with a minimum coating gap of 5 μm, as shown in Figure 1. The second was a dip coating device shown in Figure 2. A machine arm was attached to grab and lift the sample upward from a solution tank. The coating speed could vary between 0.1 - 2 cm/s. The last was a slot die coater as shown in Figure 3(a). The slot die was attached to the mount of a patch coater as shown in Figure 3(b). The coating solution was delivered by a piston pump (KD scientific, KDS 100) through the slot die, and then coated on the substrate which was fixed on the marble platform of the patch coater.

Measurements

All the test solutions were coated on the polyethylene terephthalate (PET) films for analysis. A base coat was necessary to prevent the aggregation of SWCNT solution [21]. The PET film was cut into a rectangular shape, 10 cm × 15 cm. All physical properties were measured at fixed positions on the films as marked in Figure 4. The coated samples were placed in an oven and heated at 90°C for 5 minutes. Two major properties of the ovendried samples were measured. A four-point probe (MCPT600) was used to detect the sheet resistance of the samples, andan UV-visible spectrophotometer (Varian Cary 50 conc) was used for transparency measurements. The uncoated PET film was used as the reference for comparison. Standard processes were taken to detect the distributions of CNT on the PET films by the scanning electronic microscope (JEOL JSM-5600).

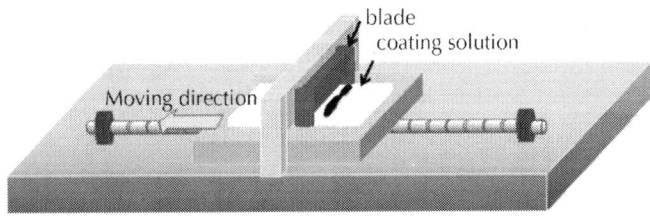

Figure 1: Blade coating operation.

Figure 2: Photo of dip coater.

Figure 3: Photos of (a) the experimental slot die with a shim; (b) the patch coater.

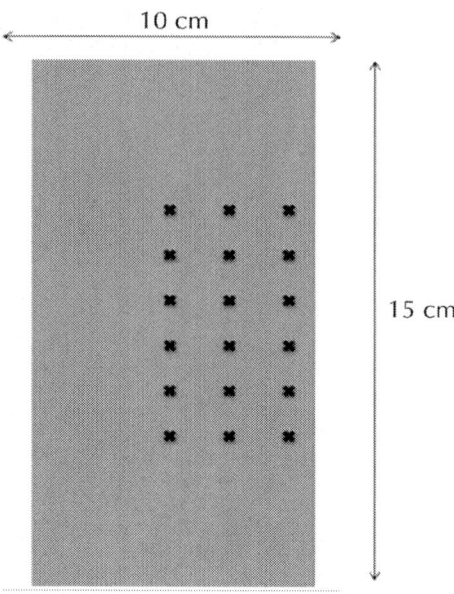

Figure 4: Sample dimensions and marked position for property measurements (10 × 15 cm²).

RESULTS AND DISCUSSION

The SWCNT solutions were coated on the PET substrates by the three coating methods. The transparency and sheet resistance of each coated sample were measured and analyzed.

The results obtained on slot die coating are presented first. In order to produce a very thin wet thickness on the slot die coater, the concentration of the SWCNT solutions must be reduced to 0.1%. This yielded a dry film with thickness as low as 5 nm. The dry film thickness t reported here is an average value which depends on flow rate, coating speed and solid content, t can be evaluated with the following formula:

$$t = \frac{Q}{V \cdot W} \cdot S\%$$

$$(1)$$

here Q is the volumetric flow rate, V is the coating speed, W is the coated width and S% is the solid content. Figure 5 presents the results of transparency and sheet resistance as a function of dry film thickness at the coating speed 10 cm/s. The results indicate that both the sheet resistance and transparency decrease as the dry film thickness increases. The increase in dry thickness is due primarily to the increasing amount of CNT accumulated on the PET substrate. Hence, it is expected that both the sheet resistance and transparency will decrease.

The effects of the coating speed on the sheet resistance and transparency for two different dry film thicknesses, 5 nm and 10 nm, are displayed in Figure 6. It is seen that both transparency and sheet resistance are independent of the coating speed for the thinner film; whereas these two properties decrease slightly as the coating speed increases for the thicker film. The sheet resistance stays around 2000 ohm/sq, and provides a transparency of around 95% for the 5 nm film; where the resistance drops to below 1000 ohm/sq, transparency reduces to about 92% for the 10 nm film when the coating velocity increases from 6 to 10 cm/s. At low coating speed, the coating solution emanating from the slot die exit will expand laterally, but the coating width will contract as the coating speed increases. The lateral movement of coating solution changes the CNT distribution, and affects the two properties. It is noted that the lateral expansion ceases to exist at high coating speed.

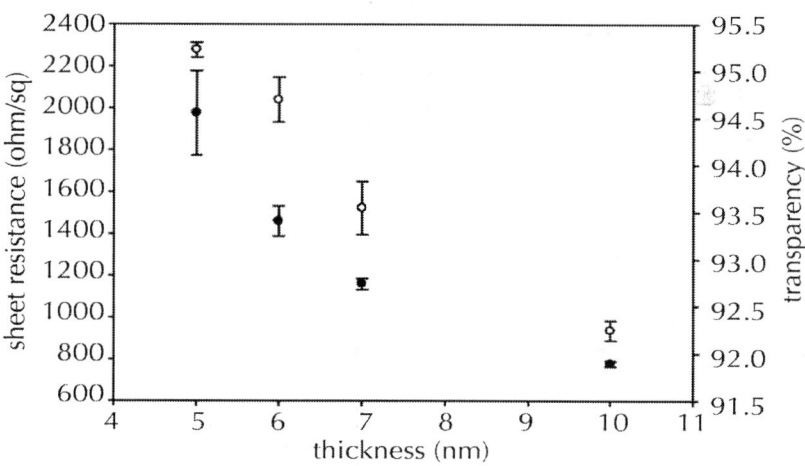

Figure 5: Sheet resistance and transparency as a function of dry film thickness for samples made by slot die coating (●: sheet resistance (ohm/sq); ○: transparency (%)).

Figure 7 shows the results obtained with a blade coater having a coating gap of 20 μm. The wet film thickness for blade coating is usually less than 50% of the blade gap for dilute Newtonian solutions [22]. The average dry film thickness can be evaluated if the solid content is known, which is around 7 - 8 nm in present study. It can be seen that both the sheet resistance and transparency decrease markedly as the coating speed increases. The transparency drops from 95% to 91%, and the resistance decreases from 3000 to below 1000 ohm/ sq when the coating speed increases from 1.0 to 6.0 cm/s. This is due to the increase of the coating thickness as the coating speed increases for a blade coater. The present observation is consistent with the conclusion of Yang and Jiang [23]. During the blade coating operation, the coating solution moves in the transverse direction as the blade advances. This lateral movement becomes more significant when the blade speed is faster.

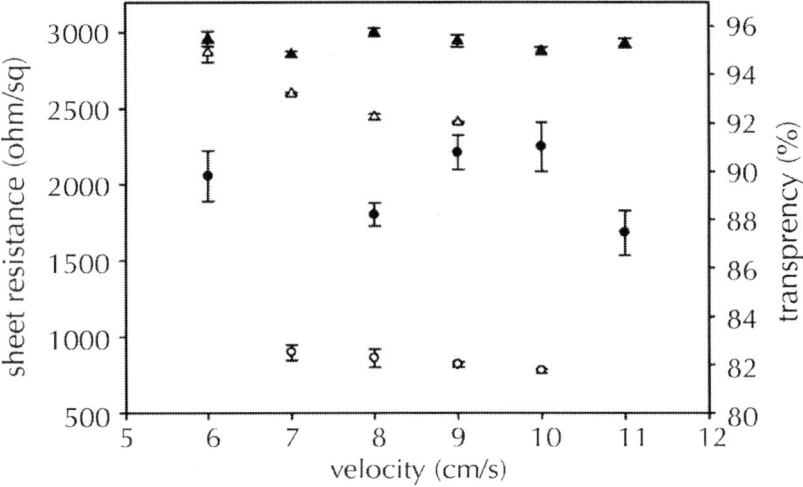

Figure 6: Sheet resistance and transparency as a function of coating speed for slot die coating (●: 5 nm (ohm/sq); ○: 10 nm (ohm/sq); ▲: 5 nm (%); Δ: 10 nm (%)).

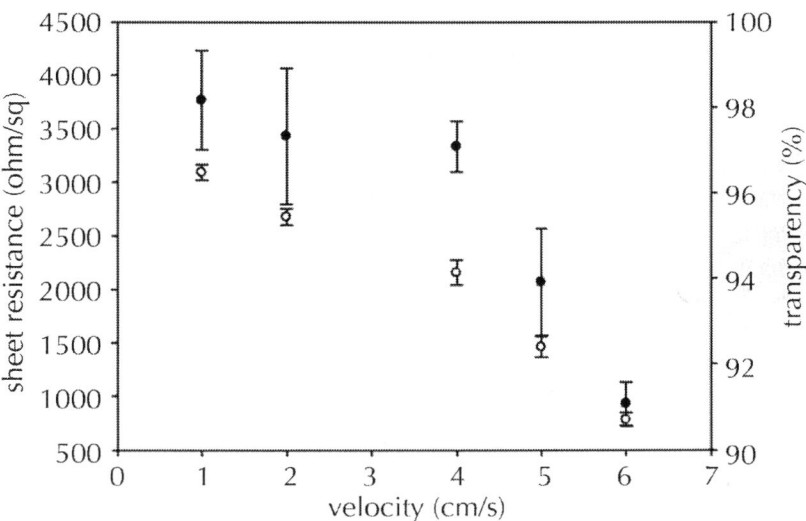

Figure 7: Sheet resistance and transparency as a function of coating speed for blade coating (•: sheet resistance (ohm/sq); ○: transparency (%)).

For a dip coating operation, the wet coated thickness is known to be proportional to the capillary number Ca as follows [24]:

$$T \equiv t_w \left(\frac{\rho g}{\mu v} \right)^{\frac{1}{2}}$$

(2a)

$$T = 0.944 (Ca)^{\frac{1}{6}}$$

(2b)

Here

$$Ca \equiv \frac{\mu v}{\sigma} \quad Ca \ll 1$$

(2c)

Here t_w is the wet film thickness, is fluid density, g is the gravitational factor, μ is the fluid viscosity and is the surface tension.

It was estimated the dry film thickness for dip coating is around 5 - 15 nm. Increasing the coating speed will increase Ca, and hence the coated film thickness for a coating solution of constant viscosity and surface tension. Therefore, both transparency and sheet resistance will drop in dip coating operated at higher velocity. The results obtained from dip coating are not presented here in similar plots as in Figures 6 and 7. Instead, they are combined for comparison purposes with the results of slot and blade coatings in the subsequent figures.

Comparison of the performance of samples based on three different coating methods is displayed inFigure 8. For dip and blade coating operations, both sheet resistance and transparency decrease with increasing coating speed. However, in order to meet the requirement of 85% transparency and sheet resistance 1000 ohm/sq, the upper limit of the coating speed for dip coating is only around 2 cm/s, and slightly above 6 cm/s for blade coating. As mentioned earlier, the optic-electronic properties appear to be independent of the coating speed in slot die coating for the 5 nm film and drop slightly for the 10 nm film when operated at a speed as high as 10 cm/s.

An attempt is made to check if a correlation exists between the sheet resistance and the transparency data obtained with the three coating methods, as shown in Figure 9. It is noted that the specific requirements of sheet resistance of around 1000 ohm/sq and transparency above 85% are satisfied for all samples obtained using the three coating methods, except for the dip coating operated at a speed of around 2 cm/s. Figure 9 reveals that there exists a linear correlation between sheet resistance and transparency. Samples with higher transparency exhibit higher sheet resistance.

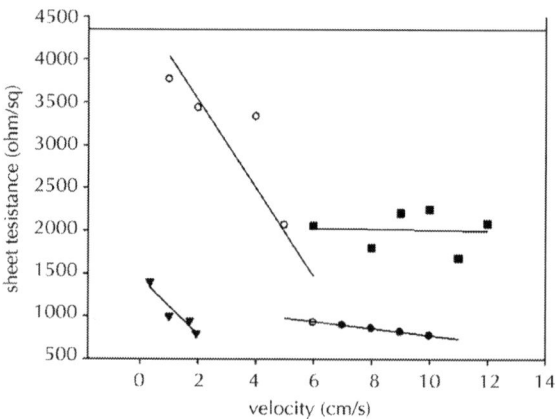

(a)

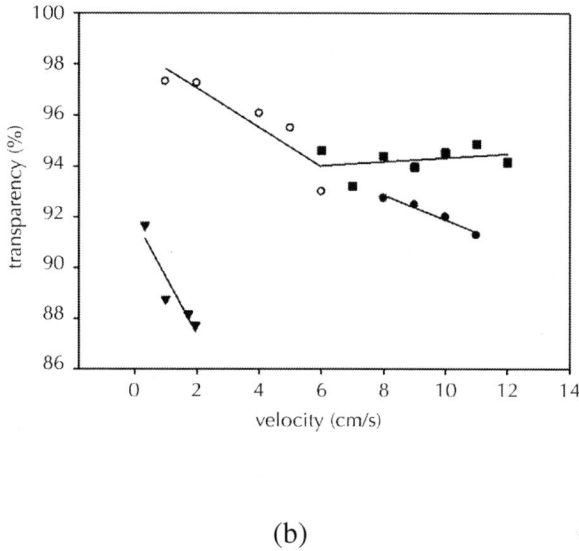

(b)

Figure 8: (a) Comparison of sheet resistance as a function of coating speed for different coating methods (■: slot 5 nm; ●: slot 10 nm; ○: blade; ▼: dip); (b) Comparison of transparency as a function of coating speed for different coating methods (■: slot 5 nm; ●: slot 10 nm; ○: blade; ▼: dip).

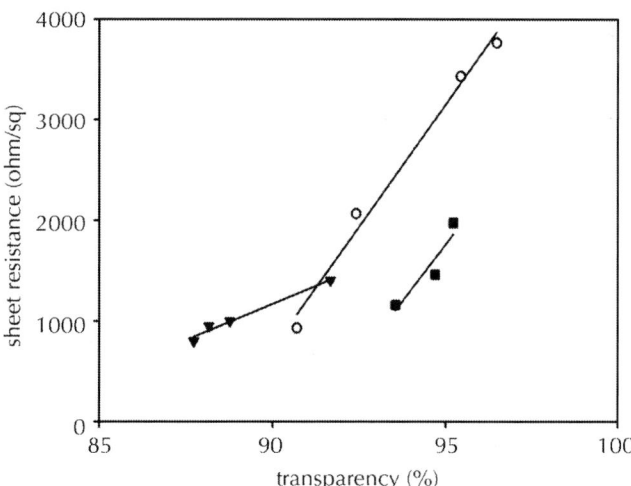

Figure 9: Sheet resistance vs. transparency for the three coating methods (■: slot 5 - 7 nm; ○: blade; ▼: dip).

The data presented in Figures 8 and 9 are based on the average value of sheet resistance and transparency. In addition, the variations of the two properties over the width of the sample are also critical to product quality. The distributions of sheet resistance for three samples are represented by the images shown in Figure 10. Figures 10(a) and (b) are two samples of different thickness obtained in slot die coating. For the thinner sample with dry film thickness of around 5 nm, the sheet resistance is quite uniform as shown in Figure 10(a). However, once the dry film thickness is increased to around 10 nm, the result in Figure 10(b) indicates that the sheet resistance at the two lateral edges increases rapidly. This implies that there are less CNT present in these regions. This is due to the lateral expansion of the coating solution after emanating from the slot die exit, which causes the concentrations of CNT towards the edges to drop sharply. The sheet resistance distribution for blade coating shown in Figure 10(c) also indicates that the distribution is not uniform due to the lateral movement of the coating solution, with high CNT concentration appearing in the central region of the sample. Therefore, to obtain a sample with uniform CNT distribution (or sheet resistance), the lateral motion has to be minimized in two-dimensional flow.

The CNT distributions on the PET films can also be observed through SEM images. The samples were taken from the central parts of the PET films. The magnifying power was selected to be 20000×. A bar of 1.0 micrometer length was given in the bottom of each figure for comparison purposes. The surfaces of the coated samples were observed. The CNT concentrations and distributions can be analyzed qualitatively. Figures 11(a) and (b) display the images obtained for two 5 nm thick dry films using the slot die coating at coating speeds of 7 cm/s and 11 cm/s, respectively. It can be seen that CNT are randomly deposited on the PET substrate, and increasing the coating speed does not cause any significant variation of the CNT distribution. The results based on blade coating are presented in Figures 11(c) and (d). Similarly, CNT are randomly distributed on the PET film, but the CNT concentration is increased substantially when the coating speed is increased from 2 cm/s to 6 cm/s as observed in Figures 11(c) and (d). This is due to the increase of the dry film thickness in blade coating at higher speed. The results based on dip coating are shown in Figures 11(e) and (f) respectively for coating speeds of 0.7 cm/s and 2 cm/s. It is seen that CNT are also randomly distributed on the PET film, but the concentrations of CNT are much higher than those obtained from the preceding two methods, and relatively unaffected by the coating speed. It must be noted that the dry films obtained from dip coating are inherently thicker than those obtained from other coating methods, thus producing a higher CNT concentration on the PET surface. The cracks appeared in the SEM images certainly are not acceptable for industrial applications. The present analysis focuses only on the sheet resistance and transparency of the coated CNT films. No binders, adhesives on other materials were added to the CNT solutions.

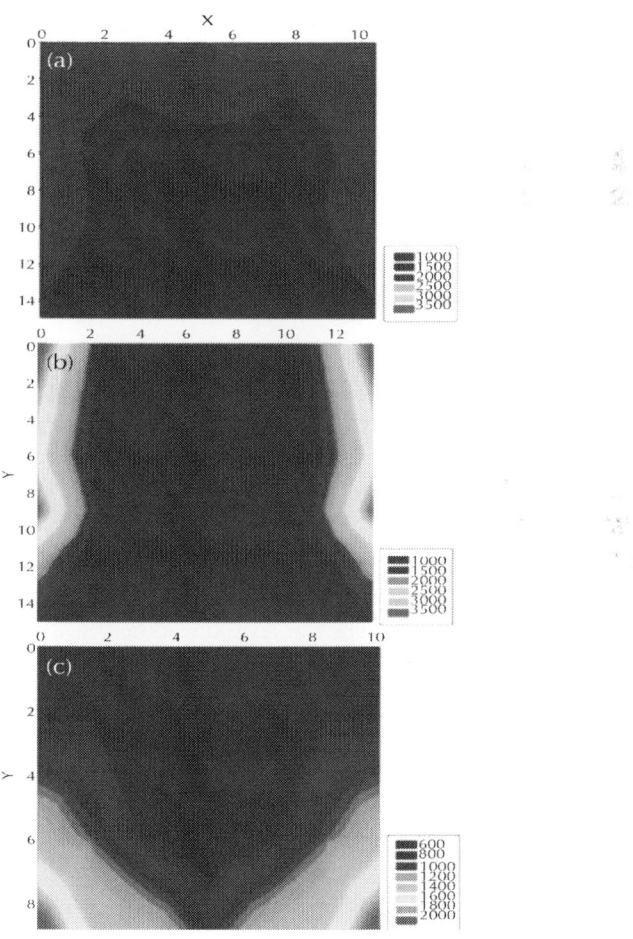

Figure 10: Images that represent sheet resistance. (a) Slot die (5 nm); (b) Slot die (10 nm); (c) Blade coating.

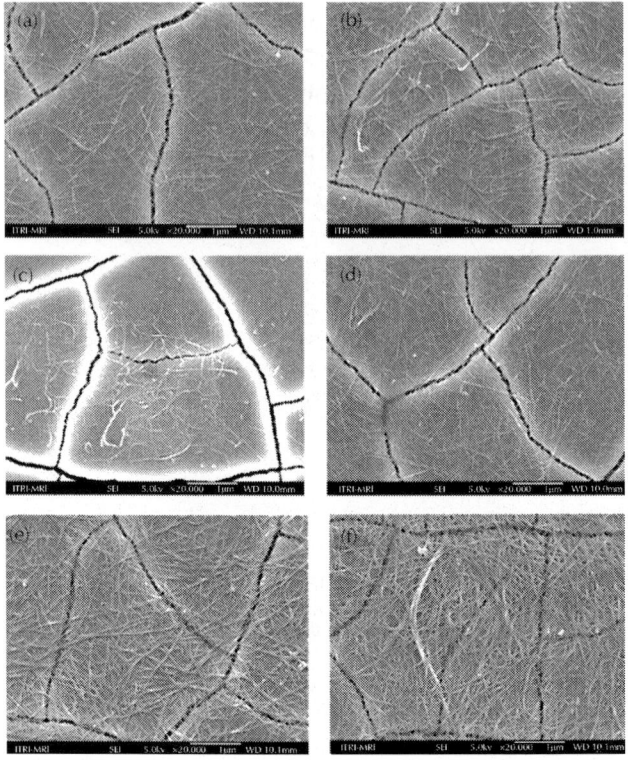

Figure 11: SEM images of samples made by the three coating methods. (a) Slot die V = 7 cm/s; (b) Slot die V = 11 cm/s; (c) Blade coating V = 2 cm/s; (d) Blade coating V = 6 cm/s; (e) Dip coating V = 0.7 cm/s; (f) Dip coating V = 2 cm/s.

CONCLUSIONS

The optical-electronicproperties of PET films coated with SWCNT obtained using three different coating methods were measured and compared. The film thickness was found to increase with increasing coating speed for blade and dip coating. The film thickness for slot die coating can be controlled provided the lateral expansion of the coated film can be minimized. Higher film thickness implies high concentration of CNT, which reduces both sheet resistance and transparency of the film.

Samples produced by the three coating methods all meet the criteria of sheet resistance 1000 ohm/sq and 85% transparency at certain coating conditions. However, at coating speed higher than 8 cm/s, only the slot die coating can produce films that meet these two criteria. The SEM images reveal that CNT were randomly distributed on PET films for samples made by the three coating methods. Distributions of sheet resistance were measured for samples obtained by slot die coating and blade coating. Coating solution may expand laterally on the PET film for blade coating and slot die coating at low speeds and high flow rates. The lateral expansion of coating solution will deteriorate the uniformity of sheet resistance. By controlling the lateral movement, slot die coating can produce samples with uniform sheet resistance.

ACKNOWLEDGEMENTS

The research work was supported by the National Science Council under Grant No. NSC99-2221-E-007-009. MY3. K.R. Chen was supported by ITRI during this study.

REFERENCES

1.   T. Minami, "New n-type Transparent Conducting Oxides," Mrs Bulletin, Vol. 25, No. 8, 2000, pp. 38-44. HHUUdoi:10.1557/mrs2000.149UU

2.   U. Ozgur, Y. I. Alivov, C. Liu, A. Teke, M. A. Reshchikov, S. Dogan, V. Avrutin, S. J. Cho and H. Morkoc, "A Comprehensive Review of ZnO Materials and Devices," Journal of Applied Physics, Vol. 98, No. 4, 2005, Article ID: 041301. HHUUdoi:10.1063/1.1992666UU

3.   D. H. Zhang and H. L. Ma, "Scattering Mechanisms of Charge Carriers in Transparent Conducting Oxide Films," Applied Physics A—Materials Science & Processing, Vol. 62, No. 5, 1996, pp. 487-492. HHUUdoi:10.1007/BF01567122UU

4.   C. C. Wang, "Deposition of Transparent Conductive Film by Wet Process," Industrial Material Magazine, Vol. 236, 2006, pp. 173-178.

5.   S. Ray, R. Banerjee, N. Basu, A. K. Batabyal and A. K. Barua, "Properties of Tin Doped Indium Oxide ThinFilms Prepared by

Magnetron Sputtering," Journal of Applied Physics, Vol. 54, No. 6, 1983, pp. 3497-3501. HHUUdoi:10.1063/1.332415UU

6.   A. Kumar and C. W. Zhou, "The Race to Replace Tin-Doped Indium Oxide: Which Material Will Win?" ACS Nano, Vol. 4, No. 1, 2010, pp. 11-14. HHUUdoi:10.1021/nn901903bUU

7.   L. B. Hu, H. S. Kim, J. Y. Lee, P. Peumans and Y. Cui, "Scalable Coating and Properties of Transparent, Flexible, Silver Nanowire Electrodes," ACS Nano, Vol. 4, No. 5, 2010, pp. 2955-2963. HHUUdoi:10.1021/nn1005232UU

8.   M. Kaempgen, G. S. Duesberg and S. Roth, "Transparent Carbon Nanotube Coatings," Applied Surface Science, Vol. 252, No. 2, 2005, pp. 425-429. HHUUdoi:10.1016/j.apsusc.2005.01.020UU

9.   O. Hjortstam, P. Isberg, S. Söderholm and H. Dai, "Can We Achieve Ultra-Low Resistivity in Carbon NanotubeBased Metal Composites?" Applied Physics A: Materials Science & Processing, Vol. 78, No. 8, 2004, pp. 1175- 1179. HHUUdoi:10.1007/s00339-003-2424-xUU

10.   L. Hu, D. S. Hecht and G. Gruner, "Percolation in Transparent and Conducting Carbon Nanotube Networks," Nano Letters, Vol. 4, No. 12, 2004, pp. 2513-2517. HHUUdoi:10.1021/nl048435yUU

11.   M. H. A. Ng, L.T. Hartadi, H. Tan and C. H. P. Poa, "Efficient Coating of Transparent and Conductive Carbon Nanotube Thin Films on Plastic Substrates," Nanotechnology, Vol. 19, No. 20, 2008, pp. 205703-205707. HHUUdoi:10.1088/0957-4484/19/20/205703UU

12.   S. De, T. M. Higgins, P. E. Lyons, E. M. Doherty, P. N. Nirmalraj, W. J. Blau, J. J. Boland and J. N. Coleman, "Silver Nanowire Networks as Flexible, Transparent, Conducting Films: Extremely High DC to Optical Conductivity Ratios," ACS Nano, Vol. 3, No. 1, 2009, pp. 1767-1774. HHUUdoi:10.1021/nn900348cUU

13.   M. C. Hersam, "Progress towards Monodisperse SingleWalled Carbon Nanotubes," Nature Nanotechnology, Vol. 3, No. 7, 2008, pp. 387-394. HHUUdoi:10.1038/nnano.2008.135UU

14.   M. S. P. Shaffer and K. Koziol, "Polystyrene Grafted Multi-Walled Carbon Nanotubes," Chemical Communications, Vol. 18, 2002, pp. 2074-2075. HHUUdoi:10.1039/b205806pUU

15. C. Richard, F. Balavoine, P. Schultz, T. W. Ebbesen and C. Mioskowski, "Supramolecular Self-Assembly of Lipid Derivatives on Carbon Nanotubes," Science, Vol. 300, No. 5620, 2003, pp. 775-778. HHUUdoi:10.1126/science.1080848UU

16. T. W. Ebbesen, H. J. Lezec, H. Hiura, J. W. Bennett, H. F. Ghaemi and T. Thio, "Electrical Conductivity of Individual Carbon Nanotubes," Nature, Vol. 382, 1996, pp. 54- 56. HHUUdoi:10.1038/382054a0UU

17. S. Kim, J. Yim, X. Wang, D. D. C. Bradley, S. Lee and J. C. Demello, "Spinand Spray-Deposited Single-Walled Carbon-Nanotube Electrodes for Organic Solar Cells," Advanced Functional Materials, Vol. 20, No. 14, 2010, pp. 2310-2316. HHUUdoi:10.1002/adfm.200902369UU

18. M. J. de Andrade, M. D. Lima, V. Skakalova, C. P. Bergmann and S. Roth, "Electrical Properties of Transparent Carbon Nanotube Networks Prepared through Different Techniques," Physica Status Solidi-Rapid Research Letters, Vol. 1, No. 5, 2007, pp. 178-180. HHUUdoi:10.1002/pssr.200701086UU

19. L. J. Ci, Y. H. Li, B. Q. Wei, J. Liang, C. L. Xu and D. H. Wu, "Preparation of Carbon Nanofibers by the Floating Catalyst Method," Carbon, Vol. 38, No. 14, 2000, pp. 1933-1937. HHUUdoi:10.1016/S0008-6223(00)00030-0UU

20. J. M. Moon, K. H. An, Y. H. Lee, Y. S. Park, D. J. Bae and G. S. Park, "High-Yield Purification Process of Singlewalled Carbon Nanotubes," The Journal of Physical Chemistry B, Vol. 105, No. 24, 2001, pp. 5677-5681. HHUUdoi:10.1021/jp0102365UU

21. S. L. Kuo, S. J. Huang and C. M. Hu, "Transparent Conductive Film and Method for Manufacturing the Same," Patent No. US20100040869-A1, 2010.

22. T. M. Sullivan and S. Middleman, "Film Thickness in Blade Coating of Viscous and Viscoelastic Liquids," Journal of Non-Newtonian Fluid Mechanics, Vol. 21, No. 1, 1986, pp. 13-38.

23. H. T. Yang and P. Jiang, "Large-Scale Colloidal SelfAssembly by Doctor Blade Coating," Langmuir, Vol. 26, No. 16, 2010, pp. 13173-13182. HHUUdoi:10.1021/la101721vUU

24. L. Landau and B. Levich, "Dragging of a Liquid by a Moving Plate," Acta Physicochimica URSS, Vol. 17, No. 42, 1942, pp. 42-54.

Ceramic Tiles Obtained from Clay Mixtures with the Addition of Diverse Metallurgical Wastes

Nancy Quaranta[1], Marta Caligaris[1], Miguel Unsen[1], Hugo López[1], Gisela Pelozo[1], Juan Pasquini[1], and Carlos Vieira[2]

[1]Grupo de Estudios Ambientales, Facultad Regional San Nicolás, Universidad Tecnológica Nacional, San Nicolás, Argentina

[2]Laboratório de Materiais Avançados–LAMAV, Universidade Estadual do Norte Fluminense Darcy Ribeiro, CCT/UENF, Campos dos Goytacazes, Río de Janeiro, Brasil

ABSTRACT

The generation of industrial residues is unavoidable, but these materials may be recovered, redirecting them toward new production processes, rather than allocating them to the stream of discards. The aim of this paper is to study the feasibility of utilization of metallurgical wastes as raw material for tiles in the ceramic industry, using the residual materials as aggregates in clay based ceramics. The residues used are: sludge and slag from several metallurgical processes, Ruthner dust and foundry sand. Samples were obtained from mixtures of clay and each waste in various percentages, which were then heat treated. The pieces obtained were characterized using several techniques, with the aim of determining the properties of these materials in relation to the commercial requirements. A high feasibility of reuse of most of these wastes as raw material in the production of ceramic bodies has been established.

INTRODUCTION

The increase in the volume of industrial discards, the simultaneous reduction in waste disposal areas, and all the problems associated with industrial wastes, are issues that demand immediate attention.

The generation of industrial residues is unavoidable. Wastes are the last outcome of any industrial activity, but these materials may be recovered, redirecting them toward new production processes, rather than allocating them to the stream of discards.

The aim of this paper is to study the feasibility of utilization of metallurgical wastes as raw material for tiles in the ceramic industry, using the residual materials as aggregates in clay based ceramics. The residues used are: blast furnace sludge (BFS), converter steel slag (CSS), Ruthner dust (RD) and foundry sand (FS).

Diverse alternatives for the reutilization of these wastes have been analyzed. Other authors have studied the replacement of sand in concrete mixtures, up to 30%, by foundry sand [1,2], the use of foundry sand in the construction of blast furnace channels [3] and the addition of foundry sand in ceramics [4].

The converter steel slag can be used in several areas, such as fertilizer [5], as aggregates in road construction [6], in cement [7] and as raw material for ceramics [8].

Ruthner dust is used, among other applications, in the manufacture of ferrites, as colorants for concrete and for catalysts [9].

Blast furnace sludge has been studied as adsorbent of heavy metals in aqueous effluents [10,11] and as the main material of ceramic particles in filters for wastewater treatment [12].

EXPERIMENTAL

The raw materials used in this study, wastes and clay, are granular and have been characterized by diverse techniques: optical microscopy (OM), scanning electron microscopy (SEM) with X-ray electron dispersive analysis (EDS), differential and gravimetric thermal analyses (DTA-TGA) and particle size distribution, among others.

The optical observations were made with Zeiss-Axiotech equipment with a Donpisha 3CCD camera and image scanner. SEM analyses were carried out through a Phillips 515 scanning electronic microscope with an X-ray detector (EDAX-Phoenix). DTA-TGA was performed in a Shimadzu DTA-50, TGA-50 with Thermal Analyzer TA-50 WSI.

Samples were obtained from mixtures of clay and each waste in various percentages. The particle sizes used were less than 2mm. A moisture content of 8% was added to these mixtures. The samples were obtained by uniaxial pressure of 25 MPa in a rectangular mold of 40 mm x 70 mm, resulting thickness of approximately 16 mm.

These samples were then heat treated in the range 900°C - 1000°C, taking into account the phase diagrams corresponding to the mixtures compositions.

The obtained probes were characterized by diverse techniques: porosity, mechanical properties, SEM, OM, weight loss on ignition (LoI), permanent volumetric variation (PVV), among others.

The porosity of the samples was determined by the 12510 IRAM Standard.

The mechanical essays were carried out with a Cific Universal Testing Machine, 294 kN. The bending test is performed on samples with the aspect ratio established in ASTM C67-03a Standard. LoI and PVV values were determined on the bricks samples by difference of weights and volumes respectively, before and after the heating treatment.

RESULTS AND DISCUSSION

Raw Materials Characterization

The microscopic appearance of the wastes is shown in Figure 1.

The chemical composition of the studied wastes by EDS semiquantitative technique is presented inTable 1, expressed as a percentage of the present elements, regardless of carbon content. There is no C present in RD waste and in FS, CSS and BFS the C content was determined as 16.18%, 13.20% y 51%, respectively. Figure 2 presents the DTA-TGA of these residues. The analysis corresponding to FS sample shows a notable weight loss between 400°C to 550°C with an exothermic peak in DTA in this range of temperatures. This is explained as the combustion reaction of the organic compounds used as additives during the fabrication process of foundry molds. A small endothermic peak at 572°C is also observed which corresponds to the reversible allotropic transformation of quartz (c.a. 573°C) [13].

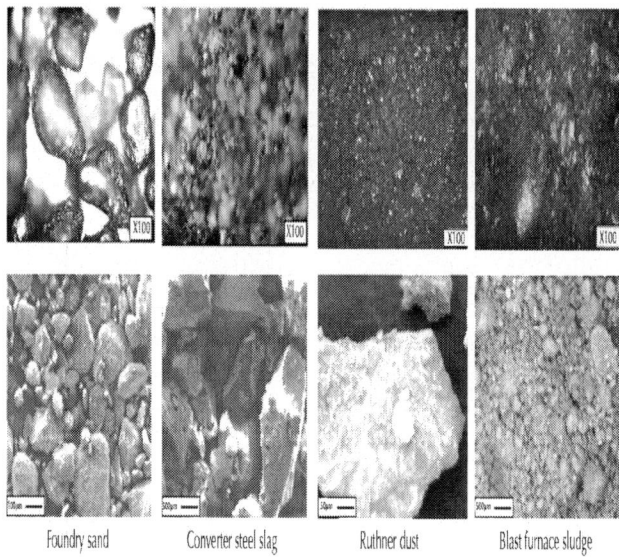

Foundry sand   Converter steel slag   Ruthner dust   Blast furnace sludge

Figure 1: Microphotographs of metallurgical wastes studied (OM photos above and SEM photos below).

Table 1: Semi-quantitative chemical analysis (EDS) of the wastes

Waste	Na	Mg	Al	Si	S	Cl	K	Ca	Mn	Fe
FS	7.1	4.9	14.7	62.8	-	-	0.5	1.1	-	8.9
CSS	-	6.8	5.4	8.1	-	-	-	53.1	3.9	22.7
RD	-	-	0.8	1.2	-	0.5	-	-	0.9	96.6
BFS	19.0	1.3	13.1	30.7	8.7	1.2	0.6	1.0	6.7	17.7

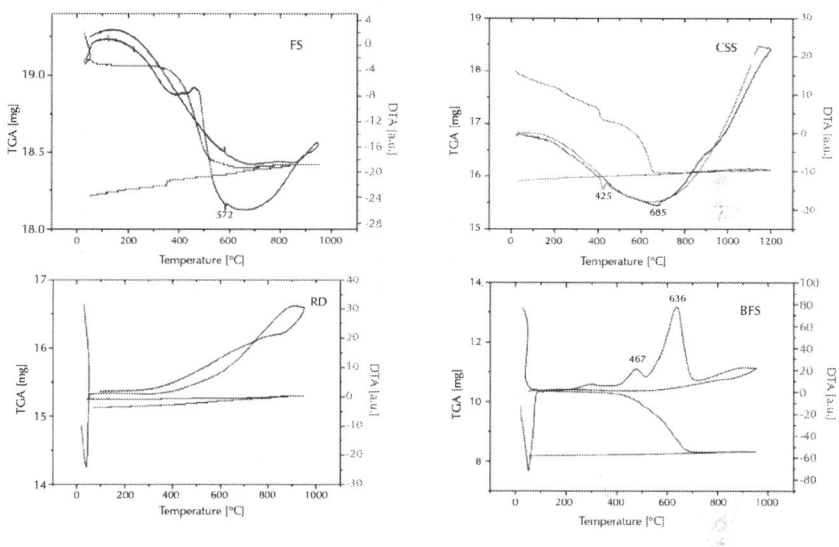

Figure 2: DTA-TGA analysis of the studies wastes.

CSS sample presents two ranges with weight losses 400°C - 430°C and 600°C - 700°C ranges, and the corresponding endothermic reactions in the DTA at maximum temperatures of 425°C and 685°C. They are interpreted as calcium hydroxide and carbonate decompositions respectively.

In BFS waste, there is a constant weight loss between 400°C to 700°C which can be attributed to the combustion of carbonaceous materials of the sample. DTA analysis shows two exothermic peaks at 467°C and 636°C. They are interpreted as exothermic oxidation reaction from Fe_3O_4 and FeO (determined by XRD) to Fe_2O_3, which is the only oxide detected by XRD in the calcined sample [14].

In RD sample, neither weight loss nor peaks have been found in DTA analysis. This result was expected taking into account the results of XRD that indicate the presence of Fe_2O_3 alone and the results of the weight loss ignition essay.

The samples were prepared with the wastes as received, selecting particle sizes less than 2 mm. The particle size distribution of the powders is shown in Figure 3.

Ceramic Products Characterization

Samples obtained with addition of CSS and BFS waste sintered at 950°C and those prepared with foundry sands sintered at 900°C.

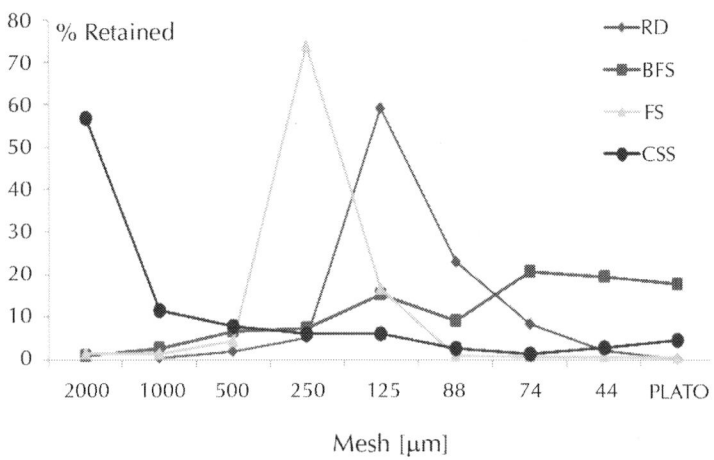

Figure 3: Particle size distribution of the materials.

Samples containing up to 20% residue RD sintered at 950°C, while those with higher RD contents did at 1000°C. Figure 4 shows some of the obtained products.

The samples are usually maintained at room temperature and humidity during the characterization process. It was noted that after a period of two weeks the samples with contents greater than 10% of CSS showed shelling of the structure and presented white areas, attributed to the formation of hydroxides of Ca and Mg, which implies that these samples are not stable even after sintering.

The porosity of the samples obtained adding BFS, FS and RD is shown in Figure 5. The corresponding results are in the required commercial range.

The porosity of the samples with CSS couldn't be determined because the samples were shelled during the corresponding test.

The test results of flexural strength are shown in Figure 6. In general these values are below those required for tiles.

Figure 4: Sintered products.

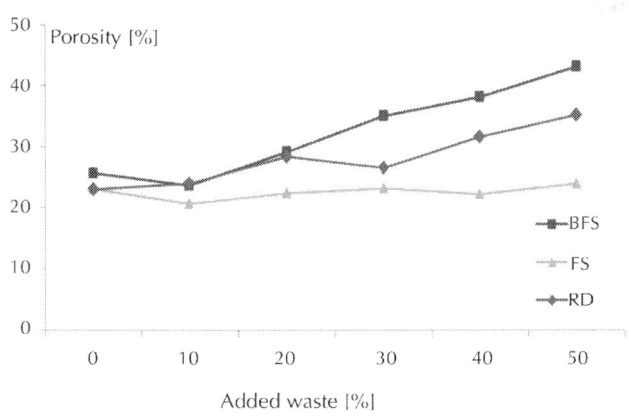

Figure 5: Porosity of the samples.

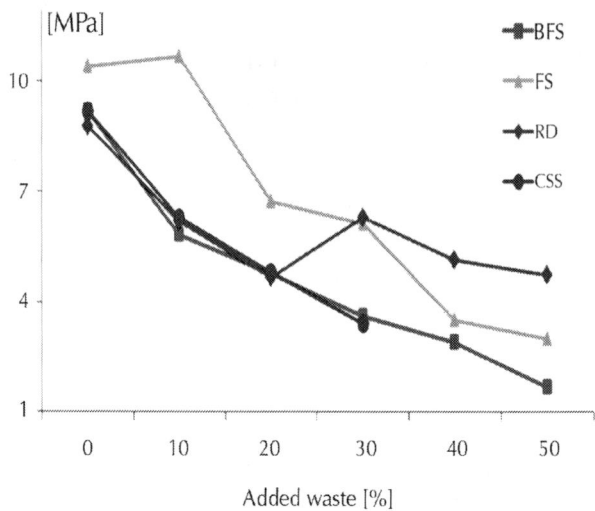

Figure 6: Flexural modulus of rupture.

The required compressive strength for building blocks, clay based, is more than 10.3 MPa, 17.2 MPa and 20.7 MPa for environmental requirements (temperature and humidity) low, moderate and severe respectively, according to ASTM C62-04. Compressive strength higher than 10.0 MPa is set by Spanish standards for this type of ceramic (NSE FL90-Royal Decree 1723/1990).

All the samples show values higher than those established by the standards, except those with higher BFS content. These results lead to the conclusion that there is a high feasibility of reuse of this type of waste as raw material for ceramic masonry, since the flexural requirements are lower.

The permanent volumetric variation (PVV) and the weight loss on ignition (LoI) of the samples are presented in Figures 7 and 8, respectively.

The PVV of the samples with added BFS is greater than that obtained for the clay except for the samples with 40 and 50% of added waste having a similar and slightly superior PVV, respectively.

The samples with lower RD contents show PVV values similar to those of the clay whereas above 20% of waste addition, variations in direct relation to discard content were determined. It is observed a maximum decrease of approximately 12% for samples with 50% of RD. Despite being a high value of volumetric variation, these samples show no cracking or visible macroscopic cracks.

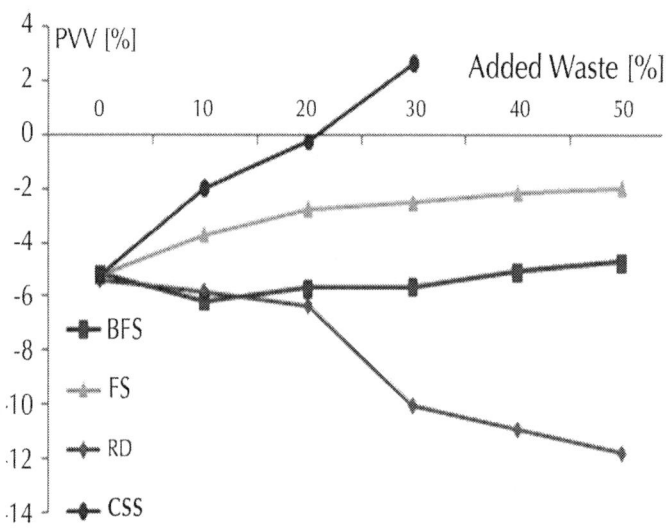

Figure 7: Permanent volumetric variation.

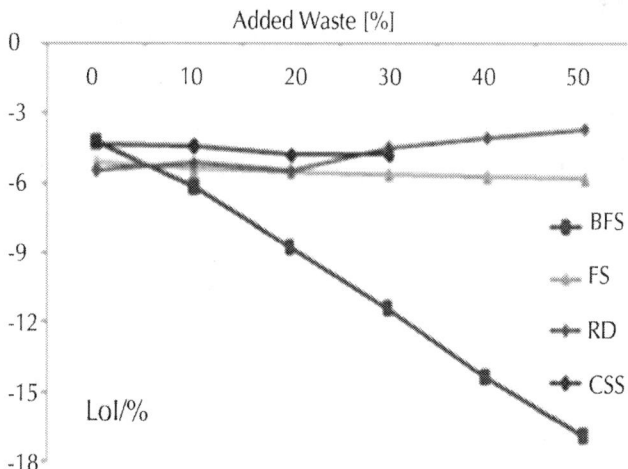

Figure 8: Weight loss on ignition of the samples.

The higher weight loss on ignition is registered for the samples with higher content of BFS waste.

The highest LoI is 16.9% for the samples with 50% BFS. The lowest LoI is registered for the sample with highest RD content.

CONCLUSIONS

This paper discusses the feasibility of utilization of metallurgical wastes as raw material for ceramic industry in tiles and masonry production. Wastes used as aggregates in clay based ceramics were blast furnace sludge, converter steel slag, Ruthner dust and foundry sand.

The results suggest that with the exception of the slag, the analyzed wastes can be incorporated as raw material replacement in the production processes of ceramic compact.

ACKNOWLEDGEMENTS

The authors wish to thank the Commission of Scientific Researches of Buenos Aires Province, Argentina, and National Agency of Technological and Scientific Promotion, Argentina, for the financial supports received for this work.

REFERENCES

1.   Z. Ismail and E. Al-hashmi, "Reuse of Waste Iron as a Partial Replacement of Sand in Concrete," Waste Management, Vol. 28, No. 11, 2008, pp. 2048-2053.http://dx.doi.org/10.1016/j.wasman.2007.07.009

2.   R. Siddique, G. Schutter and A. Noumowe, "Effect of Used Foundry Sand on the Mechanical Properties of Concrete," Construction and Building Materials, Vol. 23, No. 2, 2009, pp. 976-980. http://dx.doi.org/10.1016/j.conbuildmat.2008.05.005

3.   R. Magnani Andrade, S. Cava, S. Nascimento Silva, L. Bastos Soledade, C. Rossi, E. Leite, C. Paskocimas, J. Varela and E. Longo, "Foundry Sand Recycling in the Troughs of Blast Furnaces: A Technical Note," Journal of Materials Processing Technology,

Vol. 159, No. 1, 2005, pp. 125-134. http://dx.doi.org/10.1016/j. jmatprotec.2003.10.021

4. F. Raupp-Pereira, M. Ribeiro, A. Segadaes and J. Labrincha, "Extrusion and Property Characterisation of WasteBased Ceramic Formulations," Journal of the European Ceramic Society, Vol. 27, No. 5, 2007, pp. 2333-2340.http://dx.doi.org/10.1016/j. jeurceramsoc.2006.07.015

5. B. Das, S. Prakash, P. S. R. Reddy and V. N. Misra, "An Overview of Utilization of Slag and Sludge from Steel Industries," Conservation and Recycling, Vol. 50, No. 1, 2007, pp. 40-57. http://dx.doi. org/10.1016/j.resconrec.2006.05.008

6. J. Waligora, D. Bulteel, P. Degrugilliers, D. Damidot, J. L. Potdevin and M. Measson, "Chemical and Mineralogical Characterization of LD Converter Steel Slag: A MultiAnalytical Techniques Approach," Materials Characterization, Vol. 61, No. 1, 2010, pp. 39-48.http://dx.doi.org/10.1016/j.matchar.2009.10.004

7. Y. L. Chen, J. E. Chang, P. H. Shih, M. S. Ko, Y. K. Chang and L. C. Chiang, "Reusing Pretreated Desulfurization Slag to Improve Clinkerization and Clinker Grindability for Energy Conservation in Cement Manufacture," Journal of Environmental Management, Vol. 91, No. 9, 2010, pp. 1892-1897. http://dx.doi.org/10.1016/j. jenvman.2010.04.006

8. E. Furlani, G. Tonello and S. Maschio, "Recycling of Steel Slag and Glass Cullet from Energy Saving Lamps by Fast Firing Production of Ceramics," Waste Management, Vol. 30, No. 8-9, 2010, pp. 1714-1719. http://dx.doi.org/10.1016/j.wasman.2010.03.030

9. W. F. Kladnig, "New Development of Acid Regeneration in Steel Pickling Plants," Journal of Iron and Steel Research, Vol. 15, No. 4, 2008, pp. 1-6.http://dx.doi.org/10.1016/S1006-706X(08)60134-X

10. A. López-Delgado, C. Pérez and F. A. López, "The Influence of Carbon Content of Blast Furnace Sludges and Coke on the Adsorption of Lead Ions from Aqueous Solution," Carbon, Vol. 34, No. 3, 1996, pp. 423-426. http://dx.doi.org/10.1016/0008-6223(96)87611-1

11. A. López-Delgado, C. Pérez and F. A. López, "Sorption of Heavy Metals on Blast Furnace Sludge," Water Research, Vol.

32, No. 4, 1998, pp. 989-996.http://dx.doi.org/10.1016/S0043-1354(97)00304-7

12.   S. Li, J. Cui, Q. Zhang, J. Fu, J. Lian and C. Li, "Performance of Blast Furnace Dust Clay Sodium Silicate Ceramic Particles (BCSCP) for Brewery Wastewater Treatment in a Biological Aerated Filter," Desalination, Vol. 258, No. 1-3, 2010, pp. 12-18.http://dx.doi.org/10.1016/j.desal.2010.04.006

13.   L. P. Davila, S. H. Risbud and J. F. Shakelford, "Quartz and Silicas. Chapter 5," In: J. F. Shackelford and R. H. Doremus, Eds., Ceramic and Glass Materials, Springer, 2008, p. 73.

14.   C. F. Vieira, P. Andrade, G. Maciel, F. Vernilli and S. Monteiro, "Incorporation of Fine Steel Sludge Waste into Red Ceramic," Material Science and Engineering A, Vol. 427, No. 1-2, 2006, pp. 142-147. http://dx.doi.org/10.1016/j.msea.2006.04.040

Chapter 9

Ways of Analysis of Fire Effluents and Assessment of Toxic Hazards

Abdulrhman M. Dhabbah

Department of Forensic Science, King Fahad Security College, Riyadh, Saudi Arabia

ABSTRACT

Fire effluents, in most cases, have an adverse effect on human health and the environment. Exposure to some compounds may show both acute and chronic toxicity. There is a lack of knowledge on the effect of organic products on the human body in terms of the rate of organic material production in fires and their degree of toxicity. Thus, there is a need to expand the scope of studies about the organic products generated from fires and improve the methods of assessment to be included as part of fire hazard assessment. Different factors can be contributed to this lack of knowledge. For example, the composition of organic products generated from fires changes progressively and rapidly with progression of combustion and in a manner that is dependent on the fire condition. It is difficult to identify individual organic compounds produced during combustion. Another key factor is the lack of suitable instruments for measuring organic products generated from a fire. Also, the lack of procedures that are used to evaluate the lethal concentration limits and the lethal dose for a broad range of organic compounds generated from a fire may be another important factor which can be contributed to this lack of knowledge.

INTRODUCTION

Polymeric materials, whether natural or synthetic, play a crucial role in our life today. They are being used in daily production processes in different sectors such as construction, transport, electrical, electronic equipment, furniture, etc. Polymers have become indispensable materials, whose major disadvantage is their easy ignition and flame spread with many of toxic combustion products.

Initially, there are three types of test fires which have been used to classify fire hazards. The first is small- scale, which is used for determination of the ignitability and sometimes flame spread. The second are bench- scale tests, which are used for the ignitability of fire spread as heat release of samples between 10 g and 250 g and for generating toxic fire effluents under controlled conditions. Finally, large scale tests are used for assessing the risks associated with specific scenarios regarding building, transport and then validating data from bench-scale tests and models of scale-up [1] .

According to the International Organisation for Standardisation (ISO), fire is defined as the "process of combustion characterized by the emission of heat and fire effluents and is usually accompanied by smoke, flame, glowing or a combination thereof" [2] . Also, ISO has defined fire effluent as the "totality of gases and aerosols, including suspended particles, created by combustion or pyrolysis in a fire" [2] . Therefore, in fire atmospheres, fire effluents (toxic gases, visible smoke and heat) have a broad role negatively affecting safety of life [3] . A wide range of common materials produce a cocktail of combustion products, which have toxic effects ranging from asphyxiation and other acute toxic effects on sub-lethal or sub-acute effects [4] [5] . Fire statistics show that inhalation of fire effluents in fire is a major cause of both serious injuries and deaths from fire [6] .

CLASSIFICATION OF FIRE EFFLUENT

Fire effluents are formed in fire by either incomplete combustion or pyrolysis of materials. They can be classified into four main categories: asphyxiant (narcotic) gases, irritant gases, sub-acute toxicants and sub-lethal toxicants. The first category includes asphyxiant gases. In general, asphyxia can be defined as the deficient supply of oxygen to the human body, which prevents the oxygen uptake by cells, leading to unconsciousness and death [7] . The main asphyxiant gases include carbon monoxide (CO) and hydrogen cyanide (HCN). Carbon monoxide (CO) is a leading cause of death in fires resulting from the formation of carboxyhaemoglobin (COHb) in the blood; this complex has a stability that is 200 times more than that of oxyhaemoglobin, which prevents the transport of oxygen from the lungs to the cells in the body [8] . Hydrogen cyanide (HCN) is also reported as one of the leading causes of death. It is released when the burning materials contain nitrogen such as wool, nylon, polyurethane and polyacrylonitrile [9] . It is also worth mentioning that toxicity of hydrogen cyanide is a factor of 20 greater than that of carbon monoxide [8] .

The second category includes irritant gases. They are classified into organic and inorganic irritants depending on both the source and molecular structure of the compounds [7] . Significant organic irritants include oxygenated organics such as acrolein and formaldehyde. Inorganic irritants include nitrogen oxides, sulphur oxides and hydrogen halides [10] [11] . Hydrogen halides are released when the fire contains halogenated materials such as polyvinyl chloride, brominated flame retardants and polytetrafluoroethylene (teflon) [4] .

Exposure to irritant gases and smoke may cause breathing difficulties with severe pain in eyes, nose, throat and chest, and in some cases lead to death either by impaired vision and preventing escape, or by inhalation of particulate matters which are sufficiently small to penetrate and accumulate in the respiratory tract [8] [12] .

Two main effects occur upon exposure to irritants. First, painful sensory irritation occurs immediately in the upper respiratory tract, while lung inflammation and pulmonary oedema occur over a period of time deeper in the lung [12] [13] .

The third category of fire toxicants includes sub-acute toxicants such as carcinogens; for instance volatile organic compounds (VOCs) such as benzene, and semi-volatile organic compounds (SVOCs) include polycyclic aromatic hydrocarbons (PAHs) such as naphthalene, benzo(a)pyrene and 2-nitrophenol.

The fourth category of fire toxicants corresponds to the sub-lethal toxicants which may exhibit a range of ill effects from respiratory sensitivity (such as isocyanates) toteratogenic and mutagenic effects, such as those caused by halogenated dibenzo-p-dioxins and dibenzofurans.

As mentioned earlier, as fire studies developed, most fire deaths have happened because of inhalation of toxic gases or impairing escape capability [8] [14] . Thus, one of the important roles of fire hazard assessment is the prediction of the hazards to life of people from exposure to fire effluents. Several combustion devices have been developed for understanding the role of combustion materials on the whole process of fire hazard assessment [15] . Therefore, quantifying the toxicity of a fire effluent is a fundamental component of any fire hazard analysis.

GENERATION OF FIRE EFFLUENTS

The degree of toxicity of combustion products depends on two factors; the material formulation and fire conditions [17] . The first factor has been discussed earlier. Regarding fire conditions, there are different stages that exhibit different types and yields of fire effluents: oxidative pre-ignition (non-flaming), early/well-ventilated flaming and fully developed under-ventilated flaming. Significant parameters used in distinguishing between the stages of fire conditions include heat fluxes, oxygen concentration, equivalence ratio, combustion efficiency and composition of materials [17] [18] . These stages have been classified in the ISO-19706 standard which is intended to serve as general guidelines for the assessment of the fire threat to people [19] . Table 1 shows the most important fire stages.

Some compounds have been identified in terms of assessment of their toxicity in the whole fire effluent. According to ISO carbon dioxide (CO_2), carbon monoxide (CO), oxygen depletion (ΔO_2), hydrogen chloride (HCl), hydrogen bromide (HBr), hydrogen cyanide (HCN), hydrogen fluoride (HF), nitric oxide (NO), nitrogen dioxide (NO_2) and sulphur dioxide (SO_2) as the main analytes for assessment of fire effluent toxicity. However, this list is not exhaustive; fires usually produce a mixture of organic compounds with varying degrees of toxicity [15] .

The effects of these organic products on the human body in terms of toxicity are not fully understood. However, the materials and conditions which result in production of these toxicants are much poorly understood [8] . The reasons behind this can be attributed to several factors. First factor is the fact that the composition of organic products fire changes progressively and rapidly, depending on the fire conditions. Therefore, it is necessary to identify individual organic compounds produced during each stage of combustion. The second major factor is the lack of suitable instruments for quantifying yields of individual organic products generated in a fire. If 100 compounds are identified, the toxicity of each one needs to be multiplied by the amount detected in order to estimate which species make an important contribution to the toxicity. Although a plethora of medical committees all over the world have provided detailed tenability limits during work hours and industrial accidents. All these institutes try to cover a wide range of dangerous substances in the environment and fire. For example, a series of short term exposure limits (STELs) were provided by the Control of Substances Hazardous to Health in the UK [20] , while a list of the immediately dangerous to life or health (IDLH) was established by the National Institute for Occupational Safety and Health (NIOSH) in the USA [21] . Similarly, the acute exposure guideline limits (AEGLs) were also provided by the National Academy of Science in the USA [22] . Nonetheless, there is still limited data showing lethal concentration limits and lethal doses for a wide range of organic compounds generated in a fire. Table 2 contains a summary of available limits for assessing some organic toxic compounds.

Generally, there is need to extend the scope of studies on fire toxicity to include organic products generated from fires and improve methods of assessment to be a part of fire hazard assessment.

Table 1: ISO classification of fire stages as on ISO-19706

Fire stage	Heat (kW m−2)	Max temp (°C)		Oxygen (%)		Equivalence ratio	VCO/ V_{ro_2}	Combustion efficiency (%)
		Fuel	Smoke	In	Out			
Non-flaming								
1a. Self-sustained smouldeing	n.a.	450 - 800	25 - 85	20	0 - 20	-	0.1 - 1	50 - 90
1b. Oxidative, external radiation	-	300 - 600		20	20	-		
1c. Anaerobic external radiation	-	100 - 500		0	0	-		
Well-ventilated flaming								
2. Well-ventilated flaming	0 - 60	350 - 650	50 - 500	20	0 - 20	<0.7	<0.05	>95
Under-ventilated flaming								
3a. Low vent. room fire	0 - 30	300 - 600	50 - 500	15 - 20	5 - 10	>1.5	0.2 - 0.4	70 - 80
3b. Post-flashover	50 - 150	350 - 650	>600	<15	<5	>1.5	0.1 - 0.4	70 - 90

Table 2: Summary and comparison of potential toxicity assessment limits for some organic compounds

Toxic species	CAS number	Formula	LC50 (ppm) [7]	RD50 (ppm)) [7]	LC50/ RD50 [7]	IDLH (ppm) [21]	AEGL 10 m. (ppm) [22]			AEGL 30 m. (ppm) [22]		
							1	2	3	1	2	3
Benzene	71-43-2	C6H6	100	0.20	500	500	130	2000	-	73	1100	5600
Toluene	108-88-7	C7H8				500	200	3100	-	200	1600	6100
Xylenes	1330-20-7	C8H10				900	130	2500	-	130	1300	3600
Methanol	76-56-1	CH3OH				-	670	11, 000	-	670	4000	14,000
Acetaldehyde	70-07-0	C2H4O	20,000 - 128,000	4946	15	2000	45	340	1100	45	340	1100
Formaldehyde	50-00-0	CH2O	700 - 800	3.1	242	20	0.9	14	100	0.9	14	70
Acrolein	107-02-8	C3H5O	140 - 170	1.7	91	2	0	0.44	6.2	0	0.01	2.5
Formic acid	64-18-6	CH2O2				30	5	-	-	-	-	-
Acetic acid	64-19-7	C2H4O2				50	-	-	-	-	-	-
Crotonaldehyde	123-73-9	C4H6O	200 - 1500	-	-	50	0.19	27	44	0.19	8.9	27
Phenol	108-95-2	C6H6O	400 - 700	-	-	250	19	29	-	19	29	-

Toluene diisocyanate (TDI)	584-84-9	C9H6 N2O2			2.5	0.020	0.24	0.65	0.020	0.17	0.65
Methylene bisphenylisocyanate (MDI)	101-68-8	C15H10 N2O2			75						
Phenyl isocyanate	103-71-9	C7H5NO				-	0.012	0.036	-	0.012	0.036
Methyl isocyanate	624-83-9	C2H3NO			3	-	0.40	1.2	-	0.13	0.40
1,4-Dioxane	123-91-1	C4H8O2			500	17	580	950	17	400	950
Trans-crotonaldehyde	123-73-9	C4H6O			50	0.19	27	44	0.19	8.9	27

CHARACTERIZATION OF ORGANIC COMPOUNDS

Organic compounds are considered as the most important products in fire effluents. They include a very large and varied number of compounds. Thus, their sampling and analysis are considered to be one of the biggest problems in fire as a result of difficulty in identifying the species and knowing their impact on the human body. However, it is possible to mention some information about characterization of organic compounds as follows.

Organo Irritants

Organo irritants are formed and released in fire atmospheres through the pyrolysis under oxygen deficient conditions for materials. The production of this type of compounds is associated with large numbers of known organic materials. Depending on the type of materials, it can be estimated that organic products in fire are either innocuous or noxious. In general, the main pyrolysis products for simple hydrocarbon polymers such as polyethylene or polypropylene are innocuous. For example, no effect upon primates was found when polypropylene is pyrolysed in nitrogen. In this case, the main products are ethylene, ethane, propene, cyclopropan, formaldehyde, butane, acetaldehyde, toluene and styrene. In contrast, when these products are oxidized in air throughout the smouldering decomposition, some of them are converted to extremely irritant products. Practically, these products showed clear effects on both mice and primates which were found in an atmosphere of fire.

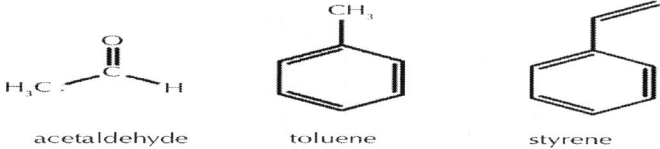

acetaldehyde   toluene   styrene

It is worth mentioning that several studies were carried out on some materials with and without fire retardant additives. The results of these experiments have confirmed that the pyrolysis of a polymer with fire retardant substances gives off the products 300 times more irritant than the pyrolysis of the same polymer without fire retardants [23] .

Polycyclic Aromatic Hydrocarbons (PAHs)

Polycyclic aromatic hydrocarbons (PAHs) are considered one of the largest groups of organic compounds; that consist of two or more joined aromatic rings. They are formed essentially as a result of incomplete combustion of organic materials through combustion atmospheres. PAHs can be formed through small unstable precursor compounds by two paths. In the first path, they are formed immediately from saturated hydrocarbons in an environment without enough oxygen. In this case, low molecular weight hydrocarbons operate as precursors in the pyro-synthesis of PAHs compounds that happen at temperatures above 500°C. The second path corresponds to thermal breakdown of heavier hydrocarbons. From these processes, several hundreds of PAHs have been investigated in most detail in the conditions of an incomplete combustion. The best known is benzo [a] pyrene (BaP), whose metabolites are mutagenic and highly carcinogenic in the body. Further, a quantity of heterocyclic aromatic compounds such as carbazole, acridine as well as nitro-PAHs can be generated.

benzo[a]pyrene (BaP)

It was proven that individual PAHs are toxic; also, they can cause a variety of non-cancer (mutagenicity, teratogenicity) and cancer effects [24] . Inhalation of PAHs is one of the main causes of cancer. For this reason, the U.S. Environmental Protection Agency (EPA) has established the toxicity equivalency factor (TEF) as a reference scale for evaluation of the risks and the toxicity of a chemical or a mixture of chemical compounds. Regarding PAHs, BaP normally is used as the reference compound which corresponds to 1.0 as TEF [24] [25].

Isocyanates

Isocyanates are a group of organic compounds characterized by their extremely reactive group (N=C=O) toward a variety of nucleophilic compounds such as alcohols, amines and water. They can be classified into three types based on the number of N=C=O groups in the molecule: monoisocyanates (one NCO) such as methylisocyanate (MIC), diisocyanates (two NCO groups) such as 2,4-toluene diisocyanate (2,4-TDI), and polyisocyanates (multiple NCOs) such as polymeric methylene diphenylisocyanate (polymeric MDI). Further, it is possible to classify isocyanates as either saturated (aliphatic and alicyclic) or aromatic (one or more aromatic rings) [26] -[28] . It is worth mentioning that diisocyanates are considered as the most important group of isocyanates, because they contain two NCO groups which allow direct polymerization reactions to occur with alcohols and formation of polyurethanes (PUR) which are used in many industrial and technical fields [26] .

4,4'-methylene diphenyldiisocyanate

Isocyanates are widely used in many syntheses, particularly for manufacturing of polymers. They are formed in any state of matter—as gases, droplets or particles. Thus, when these compounds are inhaled, various adverse health effects may happen as a result of binding them with human tissues, proteins and DNA, with production of toxic adducts and metabolites [27] [28] . There are a lot of diseases which may be caused by isocyanates including asthma, which is considered as the most common clinical diagnosis, chronic obstructive pulmonary disease (COPD), non-obstructive, extrinsic allergic alveolitis and cancer [27] [28] .

Dioxins

When incomplete combustion happens, a wide range of organic compounds are produced including dioxins which are Polychlorinated dibenzo-p-dioxins (PCDDs) and Polychlorinated dibenzofurans (PCDFs). Both these types of compounds share a common basic molecular structure which comprised two aromatic rings. Two ether-bridges join the two rings in the case of PCDFs, whereas in the case of PCDDs, they are linked by one ether- bridge and one carbon-carbon bond in a heterocyclic arrangement. Further, these compounds can be substituted by up to eight chlorine atoms; thus, there are 75 individuals of PCDDs and 135 individual of PCDFs. In contrast, the molecular structure of 2,3,7,8-tetrachlorodibenzo-p-dioxin (2,3,7,8-TCDD) is considered as the highest toxic compound in dioxins with many effects including immunotoxicity, acute dermal toxicity, reproductive health effects, teratogenicity disruption and carcinogenicity [27] -[29] .

Polychlorinated dibenzo-pioxins (PCDDs)

Polychlorinated dibenzofurans (PCDFs)

In general, there are 17 congeners of dioxin species (7 for PCDDs and 10 for PCDFs) with chlorine atoms substituted at positions 2, 3, 7 and 8. Their characteristics resemble those of 2,3,7,8-TCDD. Inhalation of dioxins is considered one of the main causes of cancer. For this reason, 2,3,7,8-TCDD is used as the reference compound for toxicity which corresponds to a value of 1.0 as TEF [27] .

Organophosphates

Organophosphates (OPs) compounds have been widely used for industrial applications such as seat cushions, pesticides and fire retardants as phosphorus fire retardants (PFRs) [30] [31] . In contrast, when exposed to OPs products, a wide range of neurotoxic effects occur by targeting the nervous system or the enzymatic system in living organisms. Several symptoms can result from exposition to OPs such as convulsions, respiratory failure, and cardiac arrhythmias [31] .

$$
\begin{array}{c}
\text{O} \\
\| \\
\text{P} \\
R^1O \diagup \;\; | \;\; \diagdown OR^3 \\
OR^2
\end{array}
$$

General structure of an organophosphate

Aerosols/Particulates

Fire-generated aerosols or particulates may cause death either by smoking which impairs escape ability by obscuring an exit route, or by producing chronic health effects and environmental hazards from chemical compounds contained in the aerosol. Particulate matters are dangerous as a result of their ability to penetrate into pulmonary tract and to pose a respiratory hazard [23] [32] . The terms and definitions of aerosol, particles and droplets were proposed as given in ISO 13943 as follows.

The first term is aerosol which is defined as a suspension of liquid droplets, or solid particles in a gas phase matrix. Furthermore, the particle sizes range from under 10 nm to over 10 μm. The second type is particles which are defined as solid-phase products present in aerosols. There are two categories of fire aerosol particles: either particles produced from incomplete combusted matter (i.e. "soot") containing a high proportion of carbon with spherical shape and about 1 μm in diameter, or particles produced from complete combusted materials consisting of small particle sized "ashes". In general, formation of soot particles is dependent on many parameters, including the types of materials, temperature, and fire conditions. The third type is droplets which are defined as liquid-phase products present in aerosols. This kind is typically generated through pyrolysis when there is not enough oxygen from both flaming and smouldering fires. Further, the condensation of water produces from combustion, around particles is possible and therefore forming aerosol droplets [32] . It is noteworthy to mention that the term particulate matter (PM) is a generic term related to an aerosol as a chemically heterogeneous material including both particles and liquid droplets found in the air.

Aerosol or particulates are generally categorised based on size:

- ultrafine particles are less than 0.1 μm in diameter;
- fine particles are between 0.1 and 2.5 μm in diameter;
- coarse particles are between 2.5 and 10 μm in diameter;
- super-coarse particles are larger than 10 μm in diameter [33] .

On the other hand, it is possible to inhale aerosols to various depths into the lungs, depending on their size and density; they can also be released into the environment and accumulate on land and in waterways.

The main biological effect of aerosols is preventing gas exchange which causes oedema and inflammation of the terminal bronchioles. Further, it is important to know that the smallest particles (<0.5 μm) can penetrate into the lungs causing interstitial and luminal oedema which is very dangerous [23] .

SAMPLING OF FIRE EFFLUENTS

Sampling is a great challenge for analysts in the fire field. It is considered the most critical part of the procedures for analysis of species and particulates in fire effluent. There are two reasons for this state of affairs. The first is that the components generated in fire atmospheres can contain a wide range of compounds and phases as well as the concentration of species in the smoke plume may change either from ppm-levels to percentage-levels or from one part of the plume to another during the course of the fire. In addition, different species have various properties. Some of them are corrosive while others are unstable or condensable vapours (including water) and aerosols which are not readily or clearly separated. The second reason is assigned to temperatures in fire atmospheres that may typically exceed 1000°C. Therefore, in most cases, gas products are hot at the sampling point which causes complications such as continued chemical reactions, or condensation of gases in different parts of the sampling device. Hence, different chemical processes may have occurred generating a mixture of novel organic molecules [15] [27] [34] .

In order to understand the effects of mixtures of toxic gases which are produced in fire effluents, it is necessary to use different methods to assess fire toxicity by having knowledge of the range, nature and concentration of species present in fire atmospheres. Several analytical methods are required to reach this goal. Therefore, a number of experimental procedures are necessary for the proper investigation of combustion products yields [6].

Ideally, the requirement for any fire effluent sampling system is to obtain a representative and realistic sample in the atmosphere tested bearing in mind that the sample should not be affected by the sampling system. Practically, this goal is achieved when a number of factors including the nature of the species, the temperature, and type of material used for the sampling probe and extract tubing, the flow rate of the sample, the type and position of both particulate filters, and condensate traps. Therefore, it is necessary to take these factors into account when sampling and analysing a fire effluent to ensure accurate identification and quantification of fire effluent products [15] [34] .

To enable the fire condition to be reliably known, for the species of interest to be generated in a fire, the steady state tube furnace SSTF (ISO DIS 19700) has proved to be reliable. It can be used to model a wide variety of fire conditions such as non-flaming, early/well-ventilated flaming and fully developed under-ventilated flaming via different combinations of temperature and different fuel/oxygen ratios. The SSTF device (ISO DIS 19700) can replicate a broad range of well-controlled combustion conditions.

Some criteria for sampling effluents are to be taken into account. For example, it is necessary to extract samples from the test atmosphere using inert sampling lines and pumps. Also, it is essential to use some suitable traps and filters, such as packed particulate traps for removing moisture before conducting the analysis, but after any heated sections of sampling tubes, to ensure that the fire effluents are free from interference before introducing them into the analysis apparatus. For this process, it is appropriate to use some agents such as glass wool to remove particulates, and calcium sulphate or calcium chloride to remove water [34] .

Sampling of Organic Compounds in Fire

Generally, a wide range of cumulative methods are available to sample and analyze organic species. These methods are various and depend on which nature of organic species is to be quantified [27] [35] . The ISO 19701: 2013 standard describes in detail all these methods [34] .

Volatile Organic Compounds (VOCs)

It is possible to measure and sample VOCs from fire gases by using commercial adsorbent tubes called Tenax. The species which can be extracted by this method include a range of non-polar or slightly polar small-medium sized hydrocarbon species with a molecular weight of approximately 75 - 200 amu. Thereafter, the sample can be analyzed by thermal desorption and then introduced to a high resolution gas chromatograph with mass spectrometric detection (GC-MS). On the other hand, commercially available activated carbon tubes can be used in the case of sampling and capturing smaller gaseous species. Relatively small sampling flows can be used for these types of commercial adsorbent tubes.

For large-scale fire gas analysis, an alternative sampling method has been developed. This method set-up consists of:

- one tube containing Amberlite XAD-2 resin to collect higher molecular weight species;
- a cooled (approximately at $-10°C$) U-tube to trap and condense water;
- one activated carbon adsorption tube cooled to $-50°C$ to collect lower molecular weight species.

In this method, greater amounts of organic compounds can be adsorbed. Furthermore, because of a relatively high sampling flow, there is no worry about risking losses in the sampling.

Both of these methods have advantages and disadvantages, as detailed below. The main advantage of the large-scale method is the possibility to perform multiple analyses on the same extracted sample. The main disadvantages of the large-scale method are:

- the possibility of some interferences occurring between the extracted species and the solvent;

- the possibility of diluting the samples below the minimum detection limit by the solvent.

- In contrast, in the case of thermal desorption, the main advantage is avoiding these effects (1 and 2), no dilution and interference can occur from a solvent, while the main disadvantage of this method is that the whole sample is consumed in a single analysis [34] [35] .

Aldehydes

Aldehydes can be sampled on cartridges consisting of silica gel covered with 2,4-dinitrophenylhydrazine (DNPH). It is worth mentioning that aldehydes have to be stabilised before analysis because of their high irritancy and their reactivity. The cartridges can be extracted with acetonitrile. Meanwhile it is possible to collect hydrazones separated by reversed-phase high performance liquid chromatography (HPLC). Samples can then be analyzed using an UV/VIS detector for formaldehyde and acetaldehyde or by atmospheric pressure chemical ionization-mass spectrometry (APCI-MS) which is preferable for all other aldehydes and ketones [34] [35] .

Semi-Volatile/Condensed Phase Organics

Large organic compounds with high boiling points, such as polycyclic aromatic hydrocarbons (PAHs) and chlorinated or brominated dibenzo-dioxins and furans (PCDDs/PCDFs and PBDDs/PBDFs), tend to be adsorbed on particles in smoke gases. Thus, it is important to use a method that enables to collect a representative sample depending on particle size distribution. For this reason, a large sampling volume is the best choice for collection semi-volatile compounds. To get a sampling speed in the orifice of the tip that is equal to that of the gas speed in the smoke gas collector duct, the diameter of the sample-probe tip should be adjusted with the sampling flow.

The sampling system should consist of:

- a heated glass fiber filter;
- a water-cooled condenser with a condensate bottle;
- a large adsorbent cartridge containing XAD-2 amberlite.

This type of sampling system is commonly used in Scandinavia.

There is another method for sampling of PAHs and PCDDs/PCDFs found in EN 1948 standard, which is in close agreement with the previous method. The difference between the two methods is that it is possible to use an adsorbent cartridge instead of using the impinger bottles which are described in the standard methods [35] [36] .

Isocyanates

Isocyanate compounds (including isocyanates, aminoisocyanates and amines) are very reactive species. These compounds can be sampled via an impinger-filter sampling system. In order to sample airborne isocyanates, it is necessary to use an impinger flask, which contains a reagent solution of di-n-butylamine(DBA) in toluene to form specific DBA-isocyanate derivatives. Furthermore, it is essential to place a glass fiber filter after the impinger. Subsequently, if the size of particles is large (>1.5 μm), they will retain in the impinger solution, whereas if the size of the particles is smaller (0.01 - 1.5 μm), they will pass all the way through the impinger solution and can be collected by the filter. It is important to bear in mind when conducting analyses of the species separately with both of the impinger solution and the filters. After that, the species can be analysed by the hyphenated technique liquid chromatography separation with mass spectroscopic detection (LC-MS) which shows a highly sensitive measurement of isocyanates, at detection limit equivalent to 0.0005 of the Swedish threshold limit value(TLV) [5] [27] [34] .

Organophosphates

For sampling organophosphates, it is necessary to pump the fire atmosphere at 0.4 L/min for 4 min to 5 min through a fritted bubbler containing 20 mL of 0.01 M sodium hydroxide, and record the total volume sampled. In order that both particulate and vapour form organophosphates can be measured, a glass fibre filter can set before the bubbler, which is then desorbed in 20 mL 0.01 M NaOH or other appropriate aqueous medium prior to analysis. However, these sampling parameters may vary from case to case if necessary to suit conditions [34] .

Aerosols/Particulates

Special techniques are required for sampling of particulates in fire effluents, which are different in some respects compared with sampling methods which are used for gases. There are four main properties which could be characteristic for both solid and liquid particulates in fire effluents:

- concentration;
- particle size distribution;
- chemical nature (which might rely on particle size);
- morphology (i.e. form and structure), which might rely on the chemical nature of particulates.

During the sampling of aerosols and particulates, it is necessary to attempt to preserve all these properties. On the other hand, the sampling probe used should be designed to operate at a velocity set to supply isokinetic sampling. This means that the velocity of sampling is similar to the velocity of the sampled effluent flow. The benefit of this process is to avoid any change in concentration or particle characteristics through the use of the probe. In addition, it is important to adjust the temperature of the probe, sampling lines, pressure and the material to get acceptable results [15] .

ANALYSIS OF FIRE EFFLUENTS

After generation of combustion products, two choices are available for sampling and analysis; either in situ for immediate analysis or extractive for subsequent analysis. In situ sampling is the method in which the chemical species are measured directly at their point of generation. Extractive sampling is the method in which the samples are collected from the fire test or experiment for analysis either immediately or at a later step. Generally, the extractive methods are more common than in situ methods [15] .

In situ analysis includes the use of non-dispersive infrared (NDIR), paramagnetism analysis and Fourier transform infrared spectroscopy (FTIR) as a direct continuous and semi-continuous analysis for determination of carbon dioxide (CO_2), carbon monoxide (CO), oxygen and a variety of volatile inorganic and organic species.

Extraction of samples may be done by the methods including:

- using a sampling line as indirect analysis then followed by gas chromatography (GC) or gas chromatography coupled to mass spectrometry (GC-MS) for analysis of many of inorganic and organic species;

- using a trapping solid adsorbent such as activated charcoal followed by gas chromatography and mass spectrometry (GC-MS) or gas chromatography equipped with flame ionization detector (GC-FID) for analysis of benzene, volatile organic compounds (VOCs) and polycyclic aromatic hydrocarbons (PAHs);

- using a trapping solution in liquid phase such as sodium hydroxide (NaOH) and hydrogen peroxide(H_2O_2) in gas-washing bottles, bubblers or impingers followed by gas chromatography and mass spectrometry (GC-MS) for analysis of the volatile organic compounds (VOCs) and polycyclic aromatic hydrocarbons (PAHs);

- using gas bags for analysis of nitrogen oxides and some of the volatile organic compounds (VOCs) [15] .

CONCLUSIONS

It is important to note that a lot of species have proven to play an essential role after considering their contribution to the toxicity of the whole fire effluent. The majority of the toxic organic compounds are formed and released in fires. In particular, they result from incomplete combustion. Emission of these products from fires, in most cases, has an adverse effect on human health and the environment. Exposure to some compounds such as volatile organic compounds (VOCs), semi-volatile organic compounds (SOVCs), polycyclic aromatic hydrocarbons (PAHs), dioxins and isocyanates may show both acute and chronic toxicity. However, this list is not exhaustive, because burning any material usually produces a complex mixture of organic compounds, with varying degrees of toxicity [15] .

The emission of organic compounds from fires, in most cases, has negative effects on the contamination of the environment and health of human. Therefore, these toxic products are most likely to cause some diseases, which include sub-acute effects and sub-lethal effects.

Moreover, there is a concern regarding their potential carcinogenicity, ill effects from respiratory sensitizes, to teratogenic and mutagenic effects, particularly for persons that are frequently exposed to these organic compounds. Furthermore, it is important to note that exposure to organic compounds during a fire can lead to fatal lung damage over periods of some hours to many days later than exposure, or permanent lung damage in survivors. The main fatal effects are lung inflammation and oedema, bronchiolitis obliterans, and bronchopneumonia [13] .

In fact, there is a lack of knowledge on the effect of organic products on the human body in terms of the rate of organic material production in fires and their degree of toxicity [8] . The reasons behind this can be attributed to several factors. The first one is the fact that the composition of organic products generated from fire changes progressively and rapidly with the progress of combustion and in a manner that is dependent on the fire condition. Therefore, it is difficult to identify individual organic compounds produced during combustion. The second major factor is the lack of suitable instruments for measuring organic products generated from a fire. A third factor is the lack of procedures that are used to evaluate the lethal concentration limits and a lethal dose for a broad range of organic compounds generated from a fire.

Thus, there is a need to expand the scope of studies about the organic products generated from fires and improve the methods of assessment to be included as part of fire hazard assessment.

ACKNOWLEDGEMENTS

The Author extends his appreciation to King Fahad Security College (Riyadh, Kingdom of Saudi Arabia) for supporting this work. He is also grateful to Professor T. Richard Hull and Dr. Anna Stec from University of Central Lancashire, School of Forensic and Investigative Sciences (Preston, United Kingdom) and Professor A. Yacine Badjah Hadj Ahmed from King Saud University, College of Science (Riyadh, Kingdom of Saudi Arabia) for their kind assistance.

REFERENCES

1.   Hull, T.R. (2010) 12-Bench-Scale Generation of Fire Effluents. In: Stec, A. and Hull, T.R., Eds., Fire Toxicity, Woodhead Publishing Ltd., Cambridge, 424-460.http://dx.doi.org/10.1533/9781845698072.4.424

2.   ISO (International Standards Orgnisation) (2008) ISO 13943.

3.   Gann, R. (2008) Fire Effluent, People, and Standards: Standardization Philosophy for the Effects of Fire Effluent on Human Tenability. Interscience, London.

4.   Alarie, Y. (2002) Toxicity of Fire Smoke. Critical Reviews in Toxicology, 32, 259-289.http://dx.doi.org/10.1080/20024091064246

5.   Gann, R., Averill, J.D., Butler, K.M., Jones, W., Mulholland, G.W., Neviaser, J.L., et al. (2001) International Study of the Sublethal Effects of Fire Smoke on Survivability and Health (SEFS) : Phase I Final Report. National Institute of Standards and Technology, Gaithersburg, Md. http://dx.doi.org/10.6028/NIST.TN.1439

6.   Stec, A.A., Hull, T.R., Purser, J.A. and Purser, D.A. (2009) Comparison of Toxic Product Yields from Bench-Scale to ISO Room. Fire Safety Journal, 44, 62-70.http://dx.doi.org/10.1016/j.firesaf.2008.03.005

7.   Purser, D.A. (2006) Assessment of Hazards to Occupants from Smoke, Toxic Gases, and Heat. In: Di Nenno, P.J., Ed., The SFPE Handbook of Fire Protection Engineering, 4th Edition, National Fire Protection Association, Quincy, 96- 193.

8.   Hull, T.R. and Stec, A. (2010) Introduction to Fire Toxicity. In: Hull, T.R. and Stec, A., Eds., Fire Toxicity, Woodhead Publishing Ltd., Cambridge, 1-25.http://dx.doi.org/10.1533/9781845698072.1.3

9.   Lestari, F. (2006) Development of in Vitro Toxicity Methods for Fire Combustion Products. Ph.D. Thesis, The University of New South Wales, Sydney, 7-71.

10.   Lemieux, P.M., Lutes, C.C. and Santoianni, D.A. (2004) Emissions of Organic Air Toxics from Open Burning: A Comprehensive Review. Progress in Energy and Combustion Science, 30, 1-32. http://dx.doi.org/10.1016/j.pecs.2003.08.001

11. Purser, D. (2012) Validation of Additive Models for Lethal Toxicity of Fire Effluent Mixtures. Polymer Degradation and Stability, 97, 2552-2561.http://dx.doi.org/10.1016/j.polymdegradstab.2012.07.009

12. Purser, D. (1992) The Evolution of Toxic Effluents in Fires and the Assessment of Toxic Hazard. Toxicology Letters, 64-65, 247-255. http://dx.doi.org/10.1016/0378-4274(92)90196-Q

13. Purser, D.A. (2010) 3—Hazaeds from Smoke and Irritants. In: Stec, A. and Hull, T.R., Eds., Fire Toxicity, Woodhead Publishing Ltd., Cambridge, 51-117.http://dx.doi.org/10.1533/9781845698072.2.51

14. Hull, T., Stec, A. and Paul, K. (2008) Hydrogen Chloride in Fires. Fire Safety Science, 9, 665-676. http://dx.doi.org/10.3801/IAFSS.FSS.9-665

15. Fardell, P. and Guillaume, E. (2010) 11—Sampling and Measurement of Toxic Fire Effluent. In: Stec, A. and Hull, T.R., Eds., Fire Toxicity, Woodhead Publishing Ltd., Cambridge, 385-423. http://dx.doi.org/10.1533/9781845698072.4.385

16. Hull, T.R., Lebek, K., Pezzani, M. and Messa, S. (2008) Comparison of Toxic Product Yields of Burning Cables in Bench and Large-Scale Experiments. Fire Safety Journal, 43, 140-150. http://dx.doi.org/10.1016/j.firesaf.2007.06.004

17. Purser, D., Stec, A. and Hull, R. (2010) 2—Fire Scenarios and Combustion Conditions. In: Stec, A. and Hull, T.R., Eds., Fire Toxicity, Woodhead Publishing Ltd., Cambridge, 26-47.http://dx.doi.org/10.1533/9781845698072.1.26

18. Hull, T.R.A., Stec, A., Lebek, K. and Price, D. (2007) Factors Affecting the Combustion Toxicity of Polymeric Materials. Polymer Degradation and Stability, 92, 2239-2246.http://dx.doi.org/10.1016/j.polymdegradstab.2007.03.032

19. ISO/TS 19706 (2004) Guidelines for Assessing the Fire Threat to People.

20. EH40/2005 Workplace Exposure Limits (2011) List of Approved Workplace Exposure Limits. Health and Safety Executive, 9-26.

21. D. for I.D.T.L. or H.C. (IDLHs) (1994) Chemical Listing and Documentation of Revised IDLH Values (as of 3/1/95). NIOSH Publications and Products.

22. National Research Council of the National Academies (2014) Acute Exposure Guideline Levels, Acute Exposure Guideline Levels for Selected Airborne Chemicals. The National Academies Press, Washington DC, Vol. 17, 3-9.

23. Stec, A.A. and Hull, T.R. (2009) Chapter 17: Fire Toxicity and Its Assessment.

24. Blomqvist, P., McNamee, M.S., Andersson, P. and Lönnermark, A. (2012) Polycyclic Aromatic Hydrocarbons (PAHs) Quantified in Large-Scale Fire Experiments. Fire Technology, 48, 513-528. http://dx.doi.org/10.1007/s10694-011-0242-9

25. Lodgejr, J. (1996) 5-Organic Pollutants, Air Quality Guidelines for Europe. Environmental Science and Pollution Research, 3, 59-120. http://dx.doi.org/10.1007/BF02986808

26. Bello, D., Woskie, Ã.S.R., Streicher, R.P., Liu, Y., Stowe, M.H., Eisen, E.A., et al. (2004) Polyisocyanates in Occupational Environments : A Critical Review of Exposure Limits and Metrics. American Journal of Industrial Medicine, 46, 480-491.

27. Blomqvist, P. (2005) Emissions from Fires, Consequences for Human Safety and the Environment. Ph.D. Thesis, Department of Fire Safety Engineering, Lund Institute of Technology, Lund University, Lund.

28. Hoesbjoer, H.N., Norwegian, T. and Inspection, L. (2001) International Consensus Report on: Isocyanates—Risk Assessment and Management. Nordic Council of Ministers, 1-131.

29. Stanmore, B. (2004) The Formation of Dioxins in Combustion Systems. Combustion and Flame, 136, 398-427. http://dx.doi.org/10.1016/j.combustflame.2003.11.004

30. Van der Veen, I. and de Boer, J. (2012) Phosphorus Flame Retardants: Properties, Production, Environmental Occurrence, Toxicity and Analysis. Chemosphere, 88, 1119-1153. http://dx.doi.org/10.1016/j.chemosphere.2012.03.067

31. Russo, M.V., Avino, P., Cinelli, G. and Notardonato, I. (2012) Sampling of Organophosphorus Pesticides at Trace Levels in the Atmosphere Using XAD-2 Adsorbent and Analysis by Gas Chromatography Coupled with Nitrogen-Phos- phorus and Ion-Trap Mass Spectrometry Detectors. Analytical and Bioanalytical Chemistry, 404, 1517-1527.http://dx.doi.org/10.1007/s00216-012-6205-2

32. ISO/DIS 29904 (2013) Fire Chemistry—Generation and Measurement of Aerosols, ISO the International Organization for Standardization, Geneva.

33. Castranova, V. (2011) Factors Governing Pulmonary Response to Inhaled Particulate Matter. 3rd Edition, John Wiley & Sons, Inc., Hoboken, 793-803.http://dx.doi.org/10.1002/9781118001684. ch38

34. BS ISO 19701 (2005) Methods for Sampling and Analysis of Fire Effluents. British Standards Institution.

35. Blomqvist, P. Simonson-McNamee, M. (2010) 13—Large-Scale Generation and Characterisation of Fire Effluents. In: Stec, A. and Hull, T.R., Eds., FireToxicity, Woodhead Publishing Ltd., Cambridge, 461-514.http://dx.doi.org/10.1533/9781845698072.4.461

36. EN1948-1 (1997) Stationary Source Emissions-Determination of the Mass Concentration of PCDDs/PCDFs-Part 1: Sampling.

10

Remediation of Agro-food Industry Effluents by Biotreatment Combined with Supported TiO$_2$/H$_2$O$_2$ Solar Photocatalysis

M. Jiménez-Tototzintle[a], I. Oller[a], A. Hernández-Ramírez[b], S. Malato[a], and M.I. Maldonado[a]

[a]Plataforma Solar de Almería-CIEMAT, Ctra. de Senés Km. 4, 04200 Tabernas, Almería, Spain

[b]Universidad Autónoma de Nuevo León (UANL), Facultad de Ciencias Químicas, Pedro de Alba s/n, Cd. Universitaria, San Nicolás de los Garza, NL C.P. 66400, Mexico

ABSTRACT

Agricultural wastewater is characterized by high organic matter and traces of organic pollutants. Effluents from the agro-food industry in particular are a potentially high environmental risk which requires appropriate, comprehensive treatment. This study evaluated the treatment of real wastewater from an agro-food processing plant by an Immobilized Biomass Reactor working in batch and continuous operating modes. Organic matter oxidation and the removal of reference organic pollutants were analyzed. Finally, the efficiency of a photocatalytic tertiary treatment with supported TiO_2 for the IBR effluent containing persistent microcontaminants was also studied. After the combined treatment, imazalil and thiabendazole were completely removed and over 90% of acetamiprid had been eliminated, but only with the additional contribution of hydrogen peroxide as an electron acceptor.

INTRODUCTION

Today's technologies, population growth, poor agricultural practices and inadequate land use have created unprecedented water pollution problems. Nearly all industrial processes involve production and discharge of massive amounts of pollutants in the aquatic and terrestrial environments, which have a strong environmental impact and require specific treatment technologies for such water to be reused [1]. Agro-food industry is one of the main business activities in the Mediterranean region. The major environmental impact involves continuous discharge of effluents polluting surface and groundwater. In the province of Almería (in southern Spain), marketing of citrus fruits and vegetables is an important agro-food sector consuming huge amounts of water for food processing. A considerable part of the resulting wastewater could be treated and safely disposed of in the environment. It is usually relatively biodegradable and can be pretreated by a conventional aerobic/anoxic biological treatment. Nevertheless, those biological systems are unable to completely eliminate non-biodegradable compounds, such as surfactants and pesticides from processing fruits and vegetables, including washing. Some fungicides, such as imazalil (IMZ), thiabendazole (TBZ) or pesticides, such as acetamiprid (ACP) can be found at trace levels ($\mu g\ L^{-1}$) in the effluent from the biological treatment at a citrus processing plant [2] and [3]. In response to part of this problem, this study presents an Immobilized Biomass Reactor (IBR) as an alternative to the conventional activated sludge treatments (CAS). IBR systems have several advantages over CAS, such as smaller plants, resistance to changes in the organic feed load, long sludge age and high specific weight, and because active biomass is grown on plastic supports, washing and maintenance are easier and post-treatment settling is unnecessary. This type of biological reactor has been used successfully for the final treatment of pretreated industrial wastewater or as a biological pretreatment step [4]. Other configurations of advanced biologic reactors for the remediation of complex wastewaters have been previously faced by other authors [5].

Recalcitrant compounds detected in biological purification system effluents actually present in such agro-food industries must be removed by a tertiary treatment before the water can be reused. Advanced Oxidation Processes (AOPs) have been widely proposed as a highly efficient alternative for tertiary treatments in industrial and municipal wastewater treatment plants due to their versatility and ability to remove recalcitrant pollutants [6]. AOPs are characterized by the production of hydroxyl radicals (HO•) able to oxidize and mineralize almost any organic molecule, yielding CO_2 and inorganic ions. The attention of researchers has been focused on AOPs that can be driven by solar radiation to avoid the high operating costs of UV lamps. Heterogeneous photocatalysis with TiO_2 is usually used due to its beneficial characteristics, including physical and chemical stability, low cost and toxicity, and excellent optical properties. Sunlight can be used because it has an appropriate energy separation between its valence and conduction bands (390 nm > > 300 nm). This AOP has been proposed for eliminating pollutants contained in wastewater, such as pesticides, drugs and dyes, and for regenerating wastewater from paper mills, municipal treatment plants, etc. [7]. However, one of its main drawbacks is that used as slurry in water, it must be removed before water disposal. Therefore, the use of TiO_2 supported on a simple, cheap material is an ongoing issue in industrial TiO_2 photocatalysis development. A wide variety of supports have been previously studied such as activated carbon, stainless steel, quartz and glass beads. TiO_2 supported on glass has been found to have good mechanical stability, because under thermal treatment, oxygen bridges have been reported to be formed between hydroxyl groups, binding the TiO_2 catalyst surface and the supporting glass. Additional advantages have been observed, including porosity, low density, natural abundance and absence of toxicity [8]. Among the disadvantages, photocatalytic activity of immobilized TiO_2 films is often lower due to a lower surface area/ volume ratio and mass transfer limitations. However, the addition of hydrogen peroxide, as an electron acceptor, can increase the efficiency of the process. Hydrogen peroxide reacts with conduction band electrons to generate more hydroxyl [9]. Photocatalytic reactors for this application must transmit UV light efficiently to the immobilized TiO_2 on the glass beads. Compound Parabolic Collectors (CPCs) have been widely studied and applied for photocatalytic wastewater treatment, for their high performance, due to the use not only of direct, but also diffuse radiation. They are a good option for adequate illumination of glass beads as CPCs do not concentrate solar irradiation but distribute it uniformly around the photoreactor tube [7].

This study evaluated the effectiveness of an advanced biological treatment using an IBR combined with a pilot-plant solar tertiary treatment based on TiO_2 supported on glass beads for removal of pesticides from wastewater from a citrus processing plant. Biotreatment evaluation included microscopic monitoring of microbiological communities, and degradation measurement of reference micro-contaminants contained in the real wastewater (imazalil (IMZ), acetamiprid (ACP) and thiabendazole (TBZ)).

MATERIALS AND METHODS

Target Wastewater

Citrus industry wastewater (CIWW) was collected from an orange juice processing plant located in the province of Almería, in south-eastern Spain. The means of the parameters measured are shown in Table 1.

Table 1: Characterization of wastewater from a citrus processing plant

pH	4.6–6.8
Conductivity (mS/cm)	2.3–5.0
Turbidity (NTU)	397–719
Inorganic Carbon (mg L^{-1})	4.3–138.1
Dissolved Organic Carbon (mg L^{-1})	1186–2380

COD (mg O_2 L^{-1})	2382–4650
Nitrite (mg L^{-1})	11–113
Nitrate (mg L^{-1})	11–300
Ammonium (mg L^{-1})	0.5–11.6
Total nitrogen (mg L^{-1})	3.5–163.0
Sulfates (mg L^{-1})	593–3068
Chlorides (mg L^{-1})	285–597
Sodium (mg L^{-1})	227–704
Potassium (mg L^{-1})	28.6–56.6
Mg^{2+} (mg L^{-1})	96–160
Ca^{2+} (mg L^{-1})	112–332
Total Suspended Solids (g L^{-1})	0.8–1.5

Analytical Measurements

Turbidity was measured with a 2100 N HACH turbidimeter. Organic matter was measured as chemical oxygen demand (COD) using Merck® Spectroquant kits, and dissolved organic carbon (DOC) measured in a Shimadzu TC-TOC-TN analyzer, model TOC-V-CSN. Total dissolved nitrogen was measured in the same TC-TOC-TN analyzer coupled to a TNM-1 unit. Total suspended solids (TSS) were determined according to American Standard Methods [10]. Anions were quantified by ion chromatography using Metrohm 872 Extension Modules 1 and 2 configured for gradient analysis. Cations were determined using a Metrohm 850 Professional IC configured for isocratic analysis. The concentration of the pesticides in the mixture was measured by UPLC (Agilent 1200 series with DAD-UV detector) using a linear gradient. The mobile phase used was a mixture of solvents A (acetonitrile) and B (ultrapure water acidified with 25 mmol L^{-1} formic acid). Simultaneous mobile phase gradient program and flow rate gradient program were used: the elution started with 100% of B, which was decreased until 85% in 7 min. After, it was programed 5-min linear gradient to 50% of eluent B and 3-min linear gradient to 100% of A, with 3 min of equilibration time. The flow rate was 1 mL min^{-1}, and the injection volume was 100 µL. The UV signal for each compound was recorded at the wavelength of maximum absorption: 300, 248 and 225 nm for TBZ ($C_{10}H_7N_3S$), ACP ($C_{10}H_{11}ClN_4$) and IMZ ($C_{14}H_{14}Cl_2N_2O$), respectively. For UPLC analyses 9 mL of sample were passed through a 0.22-µm syringe filter, then 1 mL of UPLC-grade acetonitrile was also passed through the filter to extract any compound adsorbed on the filter.

Advanced Immobilized Biomass Reactor

The IBR pilot plant installed at the Plataforma Solar de Almería (Spain) consists of a 20-L IBR conic reactor, 90% of which was occupied by K1 AnoxKaldnes supports (diameter = 9.1 mm, surface area = 500 m² m⁻³, and density = 0.95 kg dm⁻³), a conic reception tank (200 L) and a clarifier tank (40 L) for receiving the treated effluent when operating in continuous mode. This pilot plant is also equipped with pH and dissolved oxygen control systems (Fig. 1). A Watson Marlow dual-head peristaltic pump was used for the continuous mode operation.

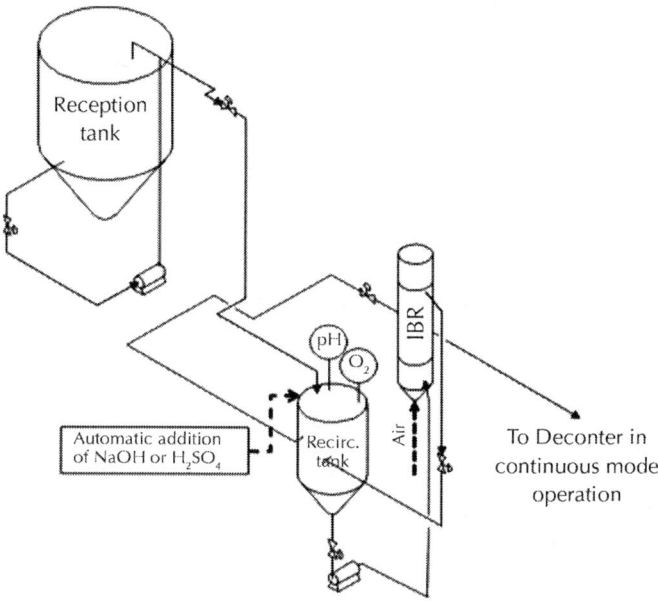

Figure 1: Schematic diagram of the IBR pilot plant.

The IBR was started up by inserting 30 L of mixed liquor from the secondary biological treatment of a municipal wastewater treatment plant (MWTP). After four days of recirculation in batch mode at 2.5 mL min⁻¹, biomass was totally immobilized on supports as confirmed by measuring TSS below the detection limit (0.01 g L⁻¹). The next step was the acclimation stage, in which diluted wastewater from the citrus plant was added (mixed with the MWTP feed stream) at concentrations from 100 mg L⁻¹ of DOC to the maximum organic load (1.2–2.4 g L⁻¹). pH was automatically maintained at 6.5–7.3 with H_2SO_4 0.1 N and NaOH 0.1 N and dissolved oxygen was always higher than 2 mg L⁻¹.

Continuous IBR operation was set up at a starting flow rate of 1 mL min⁻¹. This flow rate was raised periodically after replacement of at least twice the volume of the bio-reactor each time. Finally, the IBR effluent during continuous operation was collected for tertiary treatment for complete elimination of the remaining concentration of pesticides.

In order to measure pesticide adsorption on the biofilm, a certain amount was removed from the IBR (36.8 g) and 500 mL of acetonitrile was added to extract and dissolve pesticides that might have been adsorbed on the biofilm. This mixture was placed in an orbital shaker at 180 rpm. Support samples were taken at the beginning and end of each biotreatment. The amount of pesticides adsorbed on the biofilm per unit mass support weight, was calculated by the Eq. (1):

$$q = \frac{(C_e - C_0) * v}{m}$$

(1)

where q is the amount of compound adsorbed per mass of K1 supports, C_e is the concentration of each pesticide measured at system stationary state (after long operation and several batches), and C_0 is the starting concentration (normally zero if the biological system is adapted and operated for a specific wastewater), v (L) is the volume of acetonitrile added for extraction of contaminants (500 mL), m (g) is the mass of supports taken (36.8 g). Although usually $C_e \gg C_0$, if IBR influent conditions significantly change, C_e could decrease causing a toxic effect, and higher concentrations of contaminants would be detected in solution.

Solar Photocatalytic Tertiary Treatment based on Supported TiO$_2$

Preparation of TiO$_2$ Supported on Glass Beads

In order to coat the spherical glass beads with TiO$_2$, a thin film was prepared using the sol–gel technique. Coated glass beads were first dried at 110 °C for 90 min and later heat-treated at 400 °C for 300 min at a rate of 3 °C min^{-1}. The photocatalytic activity of TiO$_2$ supported on glass beads has been evaluated elsewhere [11].

Solar Photoreactor Pilot Plant

Solar heterogeneous photocatalytic experiments were carried out in a CPC pilot plant specially designed for solar photocatalytic applications. The photo-reactor consists of a continuously stirred tank, a centrifugal recirculation pump (mod. NH-50PX) for giving a flow of 2.5 L min^{-1}, a CPC module and connecting tubing and valves (Fig. 2) For this specific experimental device two DURAN™ borosilicate glass tubes (28.45 mm O.D) mounted on a fixed platform tilted 37° (local latitude) and uniformly packed with TiO$_2$ supported on glass beads (~4600 beads per tube) were used. The total illuminated area (A_i) was 0.250 m^2 and the total volume (V_T) was 8 L, 0.8 L of which was irradiated (V_i). Solar ultraviolet radiation (UV) was measured by a global UV radiometer (KIPP&ZONEN, model CUV 3) mounted on a platform tilted 37°. The total UV energy received per unit of volume of water (Q_{UV}) inside the reactor during the interval t can be found by Eq. (2) [12].

$$Q_{UV,n} = Q_{UV,n-1} + \Delta t_n \cdot UV_{G,n} \frac{A_i}{V_T}; \quad \Delta t_n = t_n - t_{n-1}$$

(2)

where t_n is the experimental time for each sample, $UV_{G,n}$ the average global solar ultraviolet radiation measured during t_n, and $Q_{UV,n-1}$ is the accumulated energy (per unit of volume, kJ L^{-1}) incident on the reactor for each sample taken during the experiment.

Figure 2: CPC solar pilot plant using only two borosilicate glass tubes for solar photocatalytic degradation experiment.

Photocatalytic experiments were carried out using the IBR continuous mode effluent, which was loaded in the CPC photoreactor (covered) and homogenized by turbulent recirculation in the dark for 20 min. Then collectors were uncovered and solar photocatalytic degradation started. Samples were taken every 15 min for the first hour and every 30 min thereafter.

RESULTS AND DISCUSSION

Microbiological communities were microscopically monitored from the beginning of system acclimation to the CIWW to the end of the last batch tested. In this study, protozoan and metazoan bioindicators of bioreactor performance were identified by microscopic analysis. Sample supports were taken from the IBR and re-suspended in the same water contained in the bioreactor, and finally observed under the microscope. Euglypha sp. was identified at the start of treatment (Fig. 3A). According to Fernandes et al.,Euglypha sp. indicates an operating system with nitrification, low organic load and sufficient aeration. Testate amoebae are also frequent in municipal wastewater treatment. Increase in these microorganisms is an indicator of older sludge [13]. Abundance of swimming ciliates, such as rotifers and Arcella sp., were found in the biological reactor during the CIWW treatment (Fig. 3B). Presence of Arcella sp. suggests an influent with a high biological oxygen demand. Rotifer presence reduces total biomass in suspension improving wastewater clarity [14] and [15].

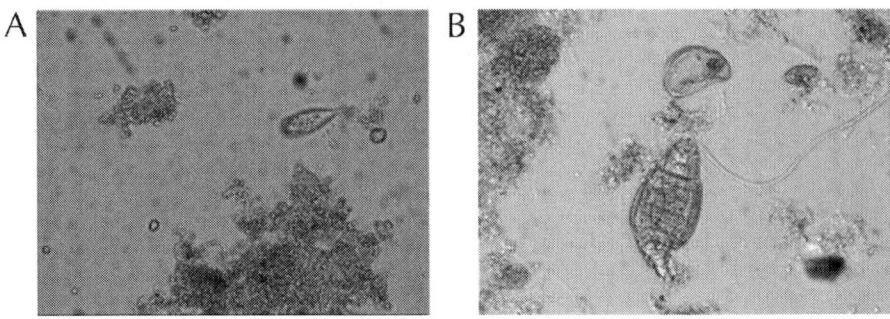

Figure 3: (A) Euglypha sp. Initial adaptation stage. (B) Rotifers and Arcella sp. found at the end of biological batch treatment. 40 × phase contrast optical microscope.

IBR Batch Operation

The IBR was operated in batch mode once the immobilized biomass became acclimated to CIWW. Fig. 4shows DOC removal and change in dissolved oxygen (DO) in the biological system during the batch mode runs. As it can be observed, DO decreased during IBR loading, and when organic matter had been reduced to 70%, DO increased once again. The concentration of DO in an aerobic process should be high enough to supply oxygen to the microorganisms in the sludge, so organic matter is degraded and ammonium is converted to nitrate. However, excess DO and air flow could affect the treatment negatively by provoking a detachment of supported biomass and reducing the system efficiency [16].

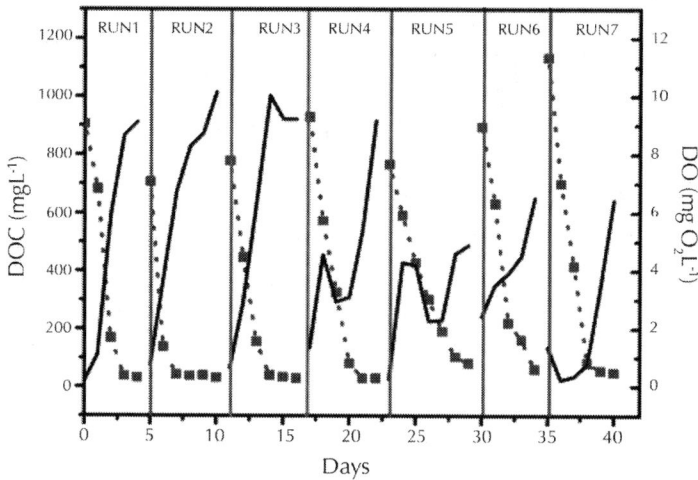

Figure 4: Change in DOC (··■··) and DO (—) during the IBR in batch mode operation.

Average oxidation in the batches run was 90–98% of organic matter after 60–72 h of treatment (Fig. 4). The maximum mineralization capacity of the biological system per volume of IBR occupied by K1 AnoxKaldnes supports was 0.44 mg of DOC L^{-1} h^{-1}. DOC removal was faster when a new batch started, but after the second day of residence time, the presence of non-biodegradable organic compounds known as the inert fraction slowed down degradation.

In the IBR system, there are anoxic micro zones in the interior of the biofilms and in the center of the reactor column due to a DO concentration gradient arising from diffusional limitations of oxygen. This contributes to nitrification and denitrification taking place in the same system [17]. As the final objective is not only the pesticide degradation monitoring in the IBR system but also the efficient performance of the process itself by also guarantying a proper nitrification, Fig. 5 shows the change in total nitrogen in the IBR effluent over time. As observed, nitrification was achieved on average of 50% after 60–72 h of treatment, improving the quality of the industrial effluent enough for a better assessment of the subsequent tertiary treatment by TiO$_2$ supported on glass beads. The maximum nitrification capacity was 0.08 mg of N-NH$_4^+$ L^{-1} h^{-1}. Nitrification started with DOC decreasing (mineralization occurred before complete nitrification), which is indicative of a slow process meaning that a longer residence time with specific operating conditions would be required for complete conversion of ammonia to nitrate (Fig. 5 shows incomplete ammonia oxidation in some runs and nitrite oxidation in others). As stated by Haseborg et al., nitrification makes ammonia reduction to nitrate via nitrite possible by reactions (3) and (4), mainly under aerobic conditions [18].

$$NH_4^+ + \frac{3}{2}O_2 \rightarrow NO_2^- + H_2O + 2H^+ \tag{3}$$

$$NO_2^- + \frac{1}{2}O_2 \rightarrow NO_3^- \tag{4}$$

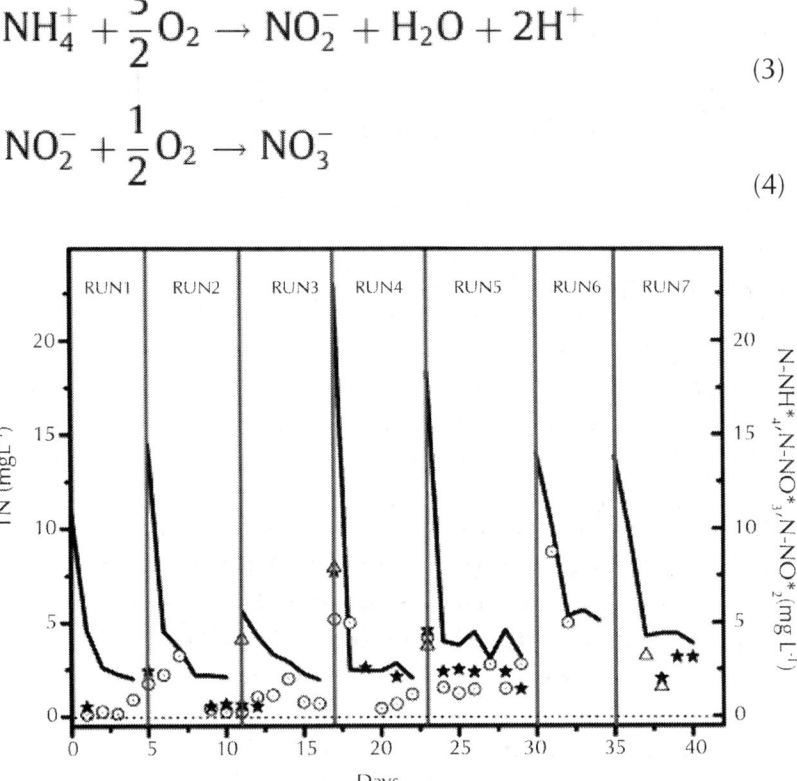

Figure 5: Change in (–) TN, (○) N-NH$_4^+$, (★) N-NO$_3^-$, (△) N-NO$_2^-$ concentrations during IBR batch mode operation.

Nitrification is performed by autotrophic ammonia-oxidizing bacteria at an optimum pH of 7.5–8.5. However, pH during IBR batch operation was always below 7.3. This effect, along with the huge organic load that must be metabolized by heterotrophic bacteria, caused inhibition of nitrification and longer reaction times were necessary for complete oxidation of ammonia and nitrite to nitrate [19].

According to nitrogen coming from NH_4^+, NO_3^- and NO_2^-, the mass balance of TN occasionally did not close properly, as observed in Fig. 5. This slight elimination of nitrogen occurred because in fixed biomass systems, dissolved oxygen on the inner wall of the biofilms on the supports is deficient due to the extremely low oxygen transfer to the inner layers of biomass. This effect, along with the preferential DO pathways inside the IBR system, which also provoke anoxic zones, made nitrate an electron acceptor leading to N_2 generation, which was eliminated into the atmosphere.

Detection of nitrite in the last run should also be mentioned. The increased TN in the influent (see Table 1 for data range) meant higher organic nitrogen content, and so, if residence time and operating conditions remained the same, reaction 3 would not take place and nitrification would be incomplete. In addition, higher ammonia, nitrite and nitrate content was linked to the increase in TN and their particular concentrations in the influent. The huge differences found in the TN content of different CIWW effluents caused such incomplete nitrification and denitrification in the biological system.

As far as we know there are a few scientific studies related to the evaluation of low concentration organic contaminants adsorption on supported biomass. However, comparing with other authors, it has been demonstrated that pesticides elimination by biological treatments always depends on their physic-chemical characteristics, as well as on the water matrix due to the presence of other organic and inorganic compounds (solvents, surfactants, etc.), which can alter the interaction of the contaminant with the activated sludge [20]. In addition, the chemical structure of the active ingredient defines the compound biodegradability and biocompatibility [21]. ACP, IMZ and TBZ degradation in the biologic reactor was monitored. Initial concentrations of each pesticide spiked in the feed stream (actual CIWW) were 90 µg L^{-1} of ACP, 70 µgL^{-1} of TBZ and 80 µg L^{-1} of IMZ. These concentrations were considered high enough to be monitored by UPLC-UV and representative of actual amounts found in that type of wastewater. Table 2 shows the percentage of degradation and the total concentration adsorbed on the biofilm of these pesticides in batch mode operation. As observed, the original concentration of pesticides gradually decreased throughout the biological treatment to 23.6% of TBZ and 93.5% of IMZ after 3 days, while ACP degradation was almost negligible. Furthermore, the partition coefficient (log K_{ow})

provides insight on the interaction of organic compounds with biomass depending on the physical–chemical properties of the pesticides, that is, pH, cation exchange capacity, ion strength and support surface area [22]. The log K_{ow} for ACP is 0.8, for TBZ 2.47 and for IMZ 3.82. Yang et al. reported that species with a log K_{ow} of 0.70–1.63 are more easily adsorbed onto the activated sludge [23]. As shown in Table 2, the order of adsorption was ACP>TBZ>IMZ. ACP, which was adsorbed on the biomass the most with a q = 0.80 µg/g support, was degraded 1%. TBZ adsorption was 27.5 µg L^{-1} (q = 0.38 µg/g of support) while IMZ adsorption was not detected according to its log K_{ow}. After this biotreatment stage, the concentrations of each reference pesticide in the IBR effluent were: 29.8 µg L^{-1} of ACP, 26.4 µg L^{-1} of TBZ and 5.2 µg L^{-1} of IMZ.

Table 2: Pesticides present in the IBR system and degradation percentage in batch mode operation

Pesticides (µg L^{-1})	Influent (µg L^{-1})	Effluent (µg L^{-1})	Adsorbed in the biofilm (µg L^{-1})	% Degradation
ACP	89.7	29.8	58.5	1.0
IMZ	80.0	5.2	0	93.5
TBZ	70.6	26.4	27.5	23.6

IBR Continuous Operation

The next step was continuous IBR operation and determination of the maximum treatment capacity of this advanced biological system. The IBR was operated continuously for 78 days. Feed flow was changed when the total IBR volume had been replaced at least twice, for more reliable results after a long enough time under the same flow conditions. Continuous mode operation started at 1 mL min^{-1} (conservative according to batch mode results) showing noticeable adaptation of the system to increasing flow rates from F4 to F8 until the treated water DOC was 20 mg L^{-1} to 50 mg L^{-1} (see Fig. 6). Fig. 7(A) shows the behavior of TN in the bioreactor at different continuous feed flow rates. Fig. 7(B) shows the percentage of TN as nitrogen from ammonia, nitrate and nitrite. As observed, bioreactor performance showed them increasing at flow rates over 4 mL min^{-1} after 69 days of biotreatment. At a flow rate of 4.4 mL min^{-1}, organic carbon elimination remained about the same, although TN increased continuously (46 mg L^{-1} after 3 days), and ammonia (38 mg L^{-1} after 3 days) even nitrite (7.0 mg L^{-1} after 3 days), showing nitrification inhibited. This was caused by two combined effects. First of all, the increase in TN in the influent due to the variable composition of the CIWW depending on the type of production resulted in incomplete nitrification, as explained above (organic nitrogen compounds are transformed into ammonia, but final oxidation to nitrate did not took place). Secondly, the increasing continuous feed flow rate enabled the system to reach maximum treatment capacity, in terms of nitrogen elimination, if operating conditions are not drastically modified by new necessities.

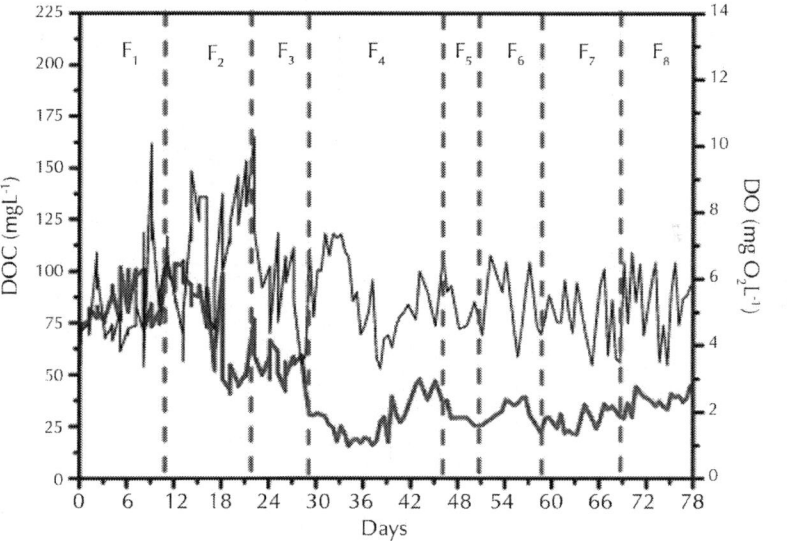

Figure 6: Residual DOC (━) in IBR effluent and change in DO (─) during continuous mode operation. F_1 = 1.0 mL min^{-1}, F_2 = 1.68 mL min^{-1}, F_3 = 2.3 mL min^{-1}, F_4 = 2.6 mL min^{-1}, F_5 = 2.8 mL min^{-1}, F_6 = 3.0 mL min^{-1}, F_7 = 4.0 mL min^{-1}, F_8 = 4.4 mL min^{-1}.

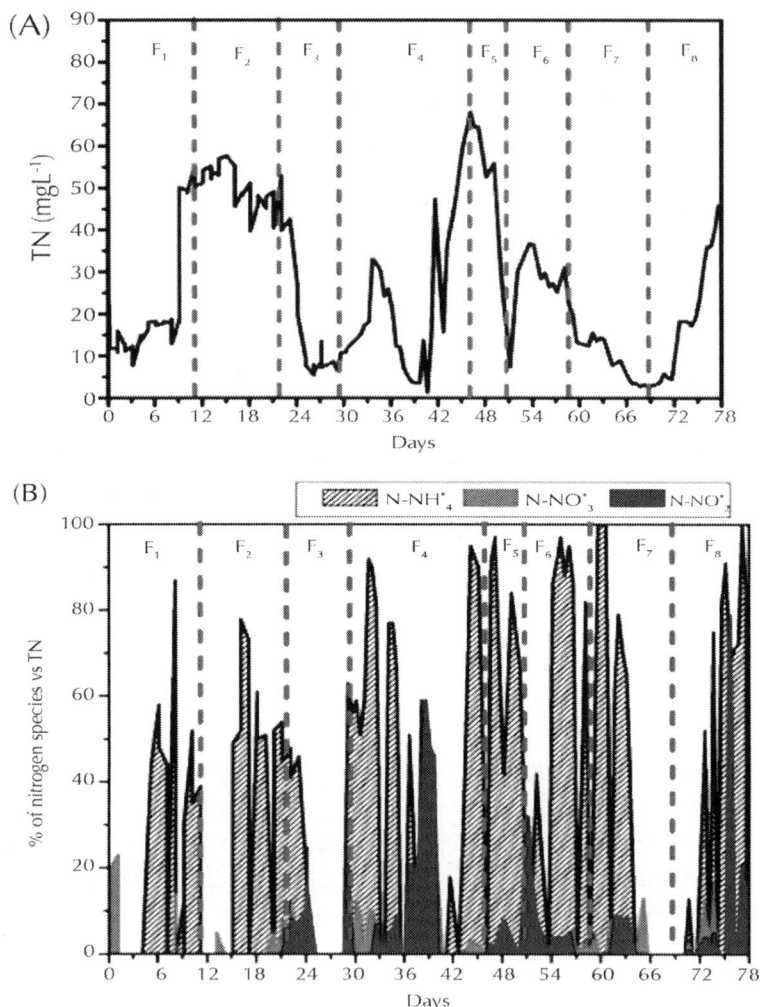

Figure 7: (A) Changes in TN in the IBR effluent during continuous operation. (B) Percentages of $N-NH_4^+$, $N-NO_3^-$ and $N-NO_2^-$ for TN measured in the IBR effluent.

The same reference pesticides as those used for IBR batch mode operation were also spiked in the continuous feed flow. During continuous mode operation, 12% and 54% of ACP and TBZ was eliminated, respectively, and IMZ degradation was 89% after 6 days of hydraulic retention time (see Table 3). In this case, pesticides were also desorbed and measured by UPLC-UV. 54.7 µg L^{-1} of ACP and 23.7 µg

L^{-1} of TBZ were adsorbed on the biofilm. Moreover, the equilibrium concentration measured per gram of supports extracted was $q = 0.13$ µg g^{-1} of support and $q = 0.07$ µg g^{-1} of support for ACP and TBZ, respectively. Again, no IMZ adsorption on the biofilms was detected.

Table 3: Pesticides present in the IBR system and degradation percentage in continuous mode operation

Pesticides (µg L^{-1})	Influent (µg L^{-1})	Effluent (µg L^{-1})	Adsorbed in the biomass (µg L^{-1})	% Degradation
ACP	85.3	20.4	54.7	12
IMZ	86.1	9.0	0	89
TBZ	78.0	12.6	23.7	54

The quality of the IBR effluent after maximum treatment capacity in continuous mode operation is summarized in Table 4. In comparison with initial values shown in Table 1 (coming from the untreated CIWW), effluent quality has been highly improved accomplishing with the physic-chemical values officially established in the Spanish regulation for industrial wastewater reuse [24]. The hydraulic retention time and high biomass concentration adsorbed on the supports not only benefited removal of bulk organic matter and elimination of nutrients, but also increased removal of organic micropollutants by increasing microbial diversity [25].

Table 4: Characterization of the IBR effluent operated under continuous mode

pH	7.2–8.0
Conductivity (mS/cm)	4.0–4.8
Turbidity (NTU)	9.0–12.0
Dissolved organic carbon (mg L^{-1})	23.2–30.23
Inorganic carbon (mg L^{-1})	5.43–6.89
COD (mg O$_2$ L^{-1})	50.4–63.0
Nitrite (mg L^{-1})	2.45–3.9
Nitrate (mg L^{-1})	3.8–4.3

Ammonium (mg L^{-1})	10.5–13.0
Total nitrogen (mg L^{-1})	15.0–17.2
Sulfates (mg L^{-1})	660–669
Chlorides (mg L^{-1})	272–284
Sodium (mg L^{-1})	554–570
Potassium (mg L^{-1})	50.32–54.3
Magnesium (mg L^{-1})	152–160
Calcium (mg L^{-1})	238–247
Total Suspended Solids (g L^{-1})	<DL

DL: Detection limit.

Solar Photocatalytic Tertiary Treatment by TiO$_2$ Supported on Glass Beads

This study proposed the use of solar photocatalysts based on supported TiO$_2$ as a tertiary treatment for complete elimination of three reference pesticides, IMZ, ACP and TBZ, to check treatment performance. The effluent from continuous operation of the IBR at its maximum flow rate was used (see Table 3 for starting pesticide concentrations). Effluent characterization is shown in Table 4. As there is an inherent decrease in surface area available for the reaction with an immobilized catalyst compared to slurry, the oxidation rate of organic carbon by HO$^{\bullet}$ is slower. When treating real wastewater the presence of other organic and inorganic species (HCO$_3^-$, SO$_4^{2-}$, Cl$^-$, etc.), which may act as scavengers of the oxidation agents (see Table 4), could also lower the reaction rate below the usual tests in demineralized water.

Fig. 8 shows the degradation profile of each reference pesticide contained in the IBR effluent by solar photocatalysis using supported TiO$_2$ on glass beads. Only IMZ degradation was complete after UV accumulated energy of 4 kJ L^{-1}, while only 46% and 22% of TBZ and ACP, respectively, were removed after 20 kJ L^{-1}. DOC (23.2 mg L^{-1}) in the IBR effluent decreased only 10%. Adsorption of ACP and TBZ on the support catalyst in the dark was found to be less than 5%, and the degradation profile observed coincided with previously published results [26].

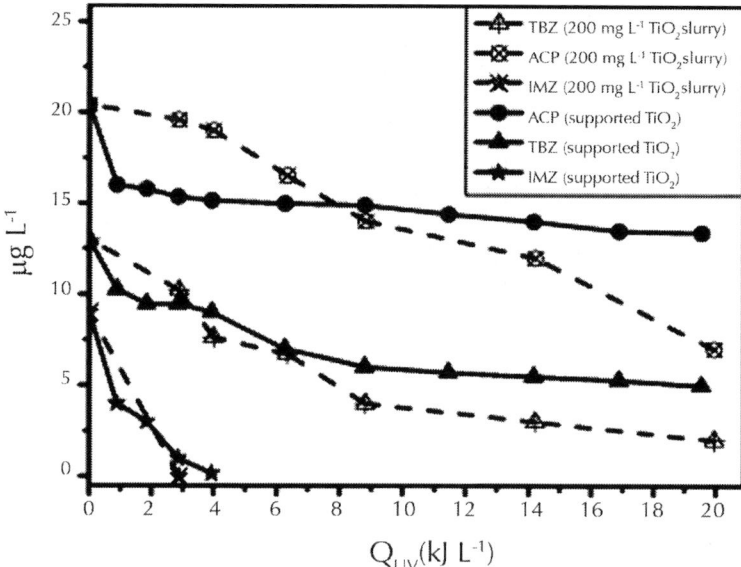

Figure 8: Comparison of the reference pesticides degradation by supported TiO$_2$ in slurry.

An important factor that must be taken into account in the degradation of ACP and TBZ, is the adsorption of those pollutants on the catalyst surface, in which case they can react with active photoinduced holes (h^+) and HO$^\bullet_{ads}$, but this interaction between the surface of the catalyst and the pollutant depends on the isoelectric point of TiO$_2$ (pzc = 6.5). The surface charge of titania is a function of the solution pH, which is affected by the reactions that occur on the particle surface as shown in Eqs. (5) and (6). Accordingly, the surface of TiO$_2$ could be either positively or negatively charged depending on pH.

$$pH < pH_{pzc} : Ti^{IV} - OH + H^+ \rightarrow Ti^{IV} - OH_2^+ \quad (5)$$

$$pH > pH_{pzc} : Ti^{IV} - OH \rightarrow Ti^{IV} - O^- + H^+ \quad (6)$$

At pH below about 6.7, which is the point of zero charge for EVONIK P-25 TiO$_2$, the TiO$_2$ surface becomes positively charged, while at pH over about 6.7 it is negatively charged. As seen in the dissociation of ACP (reaction (7)) and TBZ (reaction (8)), both compounds exist at pH

7.5 (see Table 4), ACP with a neutral charge and TBZ with a negative charge. In addition, the IBR effluent pH was always 7.2–8.0. Therefore, at near-neutral/slightly alkaline pH, ACP has a neutral charge while TBZ is negative, and the catalyst surfaces are also negatively charged, enhancing mutual repulsion. Therefore, photo-adsorption of ACP and TBZ on the catalyst surface is expected to be low. This may partly explain (along with the complex composition of the water matrix), the low photocatalytic activity observed at near-neutral/slightly alkaline pH. This effect has been observed previously in the photocatalytic treatment of other contaminants [27], [28] and [29]

$$(7)$$

$$(8)$$

Supported TiO_2 photocatalytic treatment was compared to conventional solar photocatalysis using 200 mg L^{-1} Evonik P25 TiO_2 in slurry, the optimal concentration for CPC reactors. As observed in Fig. 8, the degradation efficiency of the pesticide mixture in the continuous IBR effluent increased using TiO_2 in slurry. Removal percentages were 68% ACP and 85% TBZ after 20 kJ L^{-1} of accumulated UV energy. Those percentages were almost twice as high for TBZ and three times as high for ACP as with the solar photocatalytic treatment with supported TiO_2. IMZ elimination was below the detection limit after 2.5 kJ L^{-1} of accumulated UV energy compared to 4 kJ L^{-1} required for supported TiO_2. This positive effect was due to the larger surface area of TiO_2 in slurry, as many more TiO_2 particles can absorb radiation photons, increasing the active sites, and favoring hydroxyl radical generation as a result. In addition, the complex matrix of the real IBR effluent caused inactivation of active sites on the supported catalyst, resulting in cessation of ACP and TBZ degradation.

Following these results, it was decided to enhance degradation efficiency by using an electron acceptor, such as hydrogen peroxide. Taking into account that supported TiO_2 efficiency is lower than in slurries and that it could be seriously affected by the presence of additional DOC and inorganic species in real wastewater, the use of hydrogen peroxide to increase the reaction rate is justified.

H_2O_2 has two functions in photocatalysis. It accepts a photogenerated electron from the conduction band of the semiconductor, thereby promoting charge separation (Reaction (9)). It also forms $HO^{\bullet}$ radicals following reactions (9) and (10), while a possible H_2O_2 reaction with the parent compounds or intermediates cannot be excluded.

$$H_2O_2 + e^- \rightarrow OH^- + HO^{\bullet} \tag{9}$$

$$H_2O_2 + O_2^{\bullet-} \rightarrow OH^- + HO^{\bullet} + O_2 \tag{10}$$

Excess H_2O_2 may act as a $HO^{\bullet}$ scavenger [30]. Therefore, the positive effect of adding different starting H_2O_2 concentrations (100, 200 and 500 mg L^{-1}) to the supported TiO_2 system on degradation of the three reference pesticides, was studied using the effluent from a secondary municipal wastewater treatment plant (MWTP) as the matrix (pH = 7.8; DOC_i = 25.2 mg L^{-1}). The IBR could not be operated continuously with citrus industry wastewater due to the seasonality of production. MWTP effluent was chosen to replace the IBR effluent because inorganic and DOC contents are similar and a complete set of tests for evaluating H_2O_2 effect on the degradation could be performed with enough water of the same quality. The MWTP effluent was spiked with 20 µg L^{-1} of each reference pesticide (TBZ, ACP and IMZ). Change in the normalized concentration of pesticides over Q_{UV} at different starting concentrations of H_2O_2 is shown in Fig. 9. Differences between 100 and 200 mg L^{-1} of H_2O_2 were small, while higher degradation efficiencies were found at starting H_2O_2 concentrations up to 500 mg L^{-1}.

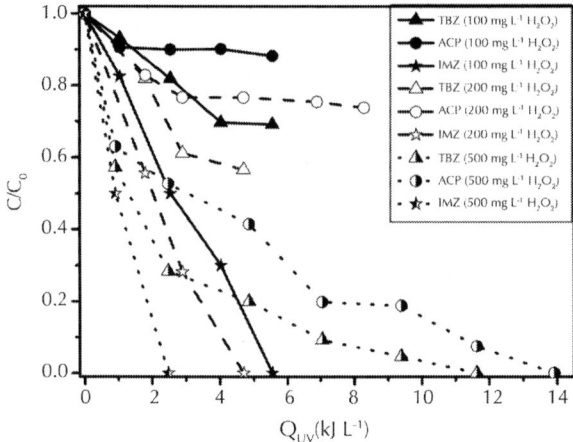

Figure 9: Photocatalytic degradation with supported TiO_2 of 20 µg L^{-1} of each reference pesticide at different hydrogen peroxide concentrations.

Finally, as the degradation results were improved by adding 500 mg L^{-1} of H_2O_2 to the supported TiO_2 system, it was also tested for direct treatment of the IBR effluent. As shown in Fig. 10, TBZ and IMZ were completely removed after 3.2 and 1.8 kJ L^{-1} of accumulative UV energy, respectively, while ACP required 15 kJ L^{-1} for 92% elimination.

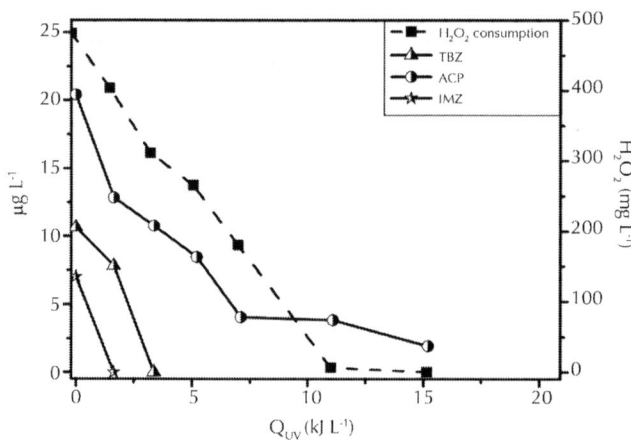

Figure 10: Degradation of reference pesticides in the IBR effluent by solar photocatalytic treatment with supported TiO_2.

Compared to the tertiary treatment of the same IBR effluent using conventional solar TiO_2 photocatalysis in slurry, these results required less energy for more degradation percentages of the reference pesticides. The complete removal of TBZ (in addition to IMZ) and the almost complete elimination of ACP at lower Q_{uv} (15 kJ L^{-1}) must be highlighted.

CONCLUSIONS

An advanced IBR biological system has been successfully tested, providing a more stable effluent from an industrial wastewater of variable composition. High removal of organic content was achieved including elimination of part of the pesticide content. Nonetheless, nitrification and denitrification were only partial due to the requirement of optimizing several operating parameters, such as pH, OD and limitation of DOC for maintaining an appropriate C/N ratio.

Biorecalcitrant compounds, such as pesticides, were not completely eliminated in the IBR system, so a final chemical oxidation step was required. The application of a polishing step based on solar photocatalysis using TiO_2 supported on glass beads was demonstrated at pilot plant scale and found to be efficient for nearly complete removal of the remaining pesticides, when H_2O_2 was added. Use of H_2O_2 for improving supported TiO_2 photocatalysis efficiency would be justified as it avoids complicated, expensive separation of TiO_2 slurries from treated wastewater.

Finally, the low accumulated UV energy necessary for such positive results in eliminating microcontaminants with solar photocatalysis should be mentioned. This is directly related to the solar collector field area required for system scale up.

ACKNOWLEDGEMENTS

The authors wish to thank the Spanish Ministry of Science and Innovation for funding under the AQUAFOTOX (CTQ2012-38754-C03-01) and HIDROPILSOL (CTQ2013-47103-R) Projects. Margarita Jiménez acknowledge the grant from the CONACYT (Consejo Nacional de Ciencia y Tecnología, Mexico).

REFERENCES

1.   I. Oller, S. Malato, J.A. Sánchez-Pérez, Combination of Advanced Oxidation Processes and biological treatments for wastewater decontamination—A review, Sci. Total Environ. 409 (2011) 4141–4166.

2.   M. Omirou, P. Dalias, C. Costa, C. Papastefanou, A. Dados, C. Ehaliotis, D.G. Karpouzas, Exploring the potential of biobeds for the depuration of pesticidecontaminated wastewaters from the citrus production chain: laboratory, column and field studies, Environ. Pollut. 166 (2012) 31–39.

3.   A. Bernabeu, R.F. Vercher, L. Santos-Juanes, P.J. Simón, C. Lardín, M.A. Martínez, J.A. Vicente, R. González, C. Llosá, A. Arques, A.M. Amat, Solar photocatalysis as a tertiary treatment to remove emerging pollutants from wastewater treatment plant effluents, Catal. Today 161 (2011) 235–240.

4.   A. Zapata, I. Oller, C. Sirtori, A. Rodríguez, J.A. Sánchez-Pérez, A. López, M. Mezcua, S. Malato, Decontamination of industrial wastewater containing pesticides by combining large-scale homogeneous solar photocatalysis and biological treatment, Chem. Eng. Technol. 160 (2010) 447–456.

5.   C.L. Amorim, A.S. Maia, R.B.R. Mesquita, A.O.S.S. Rangela, M.C.C. van Loosdrechte, M.E. Tiritan, P.M.L. Castro, Performance of aerobic granular sludge in a sequencing batch bioreactor exposed to ofloxacin, norfloxacin and ciprofloxacin, Water Res. 50 (2014) 101–113.

6.   D. Mantzavinos, E. Psillakis, Enhancement of biodegradability of industrial wastewaters by chemical oxidation pre-treatment, J. Chem. Technol. Biotechnol. 79 (2004) 431–454.

7.   S. Malato, P. Fernandez-Ibáñez, M.I. Maldonado, J. Blanco, W. Gernjak, Decontamination and disinfection of water by solar photocatalysis: recent overview and trends, decontamination and disinfection of water by solar photocatalysis: recent overview and trends, Catal. Today 147 (15) (2009) 1–59.

8.   S.N. Hosseini, S.M. Borghei, M. Vossoughi, N. Taghavinia, Immobilization of TiO2 on perlite granules for photocatalytic degradation of phenol, Appl. Catal. B Environ. 74 (1–2) (2007) 53–62.

9.   A.C. Affam, M. Chaudhuri, Degradation of pesticides chlorpyrifos, cypermethrin and chlorothalonil in aqueous solution by TiO2 photocatalysis, J. Environ. Manage. 130 (2013) 160–165.

10. L.S. Clesceri, A.E. Greenberg and A.D. Eaton, Standard Methods for the Examination of Water and Wastewater, 20th American Public (Ed.), Health Association (1998) 8–10.

11. N. Miranda-García, M.I. Maldonado, J.M. Coronado, S. Malato, Degradation study of 15 emerging contaminants at low concentration by immobilized TiO2 in a pilot plant, Catal. Today 151 (2010) 107–113.

12. L. Prieto-Rodríguez, D. Spasiano, I. Oller, F.I. Calderero, A. Agüera, S. Malato, Solar photo-Fenton optimization for the treatment of MWTP effluents containing emerging contaminants, Catal. Today 209 (2013) 188–194.

13. H. Fernandes, M.K. Jungles, K. Hoffmann, R.V. Antonio, R.H.R. Costa, Full-scale sequencing batch reactor (SBR) for domestic wastewater: performance and diversity of microbial communities, Bioresour. Technol. 132 (2013) 262–268.

14. J. Lapinski, A. Tunnacliffe, Reduction of suspended biomass in municipal wastewater using bdelloid rotifers, Water Res. 37 (2003) 2027–2034.

15. T. Sasahara, T. Ogawa, Treatment of brewery effluent. Part VIII: Protozoa and metazo found in the activated sludge process for brewery effluent, Protist, Monatsschr. Brauwiss. 36 (11) (1983) 443–448.

16. B. Holenda, E. Domokos, A. Rédey, J. Fazaka, Dissolved oxygen control of the activated sludge wastewater treatment process using model predictive control, Comput. Chem. Eng. 32 (2008) 1270–1278.

17. J.B. Holman, D.W. Wareham, COD, ammonia and dissolved oxygen time profiles in the simultaneous nitrification/denitrification process, Biochem. Eng. J. 22 (2005) 125–133.

18. E.T. Haseborg, T.M. Zamora, J. Fröhlich, F.H. Frimmel, Nitrifying microorganisms in fixed-bed biofilm reactors fed with different nitrite and ammonia concentrations, Bioresour. Technol. 101 (6) (2010) 1701–1706.

19. J.L. Campos, J.M. Garrido-Fernández, R. Méndez, J.M. Lema, Nitrification at high ammonia loading rates in an activated sludge unit, Bioresour. Technol. 68 (2) (1999) 141–148.

20. A. Peña, J.A. Rodríguez-Liébana, M.D. Mingorance, Persistence of two neonicotinoid insecticides in wastewater, and in aqueous solutions of surfactants and dissolved organic matter, Chemosphere 84 (2011) 464–470.

21. R. Rojas, J. Morillo, J. Usero, L. Delgado-Moreno, J. Gan, Enhancing soil sorption capacity of an agricultural soil by addition of three different organic wastes, Sci. Total Environ. 458 (460) (2013) 614–623.

22. R. Kode˘sová, M. Kocáreka, V. Kode˘s, O. Drábeka, J. Kozáka, K. Hejtmánkovác, Pesticide adsorption in relation to soil properties and soil type distribution in regional scale, J. Hazard. Mater. 186 (2011) 540–550.

23. S.F. Yang, C.F. Lin, A.Y.C. Lin, P.K.A. Hong, Sorption and biodegradation of sulfonamide antibiotics by activated sludge: experimental assessment using batch data obtained under aerobic conditions, Water Res. 45 (2011) 3389– 3397.

24. Guía para la Aplicación del R.D. 1620/2007 por el que se establece el Régimen Jurídico de la Reutilización de las Aguas Depuradas, ISBN: 978-84-491-0998-0, Ed. Ministerio de Medio Ambiente y Medio Rural y Marino. Spain. (2010).

25. S.K. Maeng, B.G. Choi, K.T. Lee, K.G. Song, Influences of solid retention time, nitrification and microbial activity on the attenuation of pharmaceuticals and estrogens in membrane bioreactors, Water Res. 47 (2013) 3151–3162.

26. M. Jiménez, M.I. Maldonado, E.M. Rodríguez, H.A. Ramírez, E. Saggioro, I. Carra, J.A.S. Pérez, Supported TiO2 solar photocatalysis at semi-pilot scale: degradation of pesticides found in citrus processing industry wastewater, reactivity and influence of photogenerated species, J. Chem. Technol. Biotechnol. 90 (1) (2015) 149–157.

27. N.P. Xekoukoulotakis, C. Drosoua, C. Brebou, E. Chatzisymeon, E. Hapeshib, D. Fatta-Kassinos, D. Mantzavinos, Kinetics of UV-A/TiO2 photocatalytic degradation and mineralization of the antibiotic sulfamethoxazole in aqueous matrices, Catal. Today 161 (2011) 163–168.

28. Y.Z. El-Nahhal, Development of controlled release formulations of thiabendazole, J. Agric. Food Chem. 3 (1) (2014) 1–8.

29. M. Deshmukh, C. Shripanavar, Synthesis of N0 -Carbamoyl-N-[(6-chloropyridin-3-yl) methyl] ethanimidamide, J. Chem. Pharm. Res. 3 (5) (2011) 636–637.

30. M. Kositzi, I. Poulios, S. Malato, J. Caceres, A. Campos, Solar photocatalytic treatment of synthetic municipal wastewater, Water Res. 38 (2004) 1147– 1154.

Citations

CHAPTER 1

V. Kuokkanen, T. Kuokkanen, J. Rämö and U. Lassi, "Recent Applications of Electrocoagulation in Treatment of Water and Wastewater—A Review," Green and Sustainable Chemistry, Vol. 3 No. 2, 2013, pp. 89-121. doi: 10.4236/gsc.2013.32013.

CHAPTER 2

M. Tejocote-Pérez, P. Balderas-Hernández, C. Barrera-Díaz, G. Roa-Morales and V. Bárcena, "Effect ofChytriomyces hyalinus on Industrial Wastewater Pre-Treated with Electrocoagulations in a Continuous System,"Natural Resources, doi:10.4236/nr.2012.33016.

CHAPTER 3

Henrik Hansson, Fabio Kaczala, Marcia Marques, and William Hogland, "Photo-Fenton and Fenton Oxidation of Recalcitrant Industrial Wastewater Using Nanoscale Zero-Valent Iron," International Journal of Photoenergy, vol. 2012, Article ID 531076, 11 pages, 2012. doi:10.1155/2012/531076.

CHAPTER 4

Hartmann, A., Petrov, V., Buhl, J., Rübner, K., Lindemann, M., Prinz, C. and Zimathies, A. (2014) Zeolite Synthesis under Insertion of Silica Rich Filtration Residues from Industrial Wastewater Reconditioning Advances in Chemical Engineering and Science, 4, 120-134. doi: 10.4236/aces.2014.42015.

CHAPTER 5

W. Sokół, «Optimal Aerations in the Inverse Fluidized Bed Biofilm Reactor When Used in Treatment of Industrial Wastewaters of Various Strength,» Advances in Chemical Engineering and Science, Vol. 2 No. 3, 2012, pp. 384-391. doi: 10.4236/aces.2012.23046.

CHAPTER 6

Amin, N., Ibrar, D. and Alam, S. (2014) Heavy Metals Accumulation in Soil Irrigated with Industrial Effluents of Gadoon Industrial Estate, Pakistan and Its Comparison with Fresh Water Irrigated Soil. Journal of Agricultural Chemistry and Environment, 3, 80-87. doi: 10.4236/jacen.2014.32010.

CHAPTER 7

K. Chen, H. Yeh, H. Chen, T. Liu, S. Huang, P. Wu and C. Tiu, "Optical-Electronic Properties of Carbon-Nanotubes Based Transparent

Conducting Films," Advances in Chemical Engineering and Science, Vol. 3 No. 1, 2013, pp. 105-111. doi: 10.4236/aces.2013.31013.

CHAPTER 8

Quaranta, N., Caligaris, M., Unsen, M., López, H., Pelozo, G., Pasquini, J. and Vieira, C. (2014) Ceramic Tiles Obtained from Clay Mixtures with the Addition of Diverse Metallurgical Wastes. Journal of Materials Science and Chemical Engineering, 2, 1-5. doi: 10.4236/msce.2014.22001.

CHAPTER 9

Dhabbah, A. (2015) Ways of Analysis of Fire Effluents and Assessment of Toxic Hazards. Journal of Analytical Sciences, Methods and Instrumentation, 5, 1-12. doi: 10.4236/jasmi.2015.51001.

CHAPTER 10

M. Jiménez-Tototzintle, I. Oller, A. Hernández-Ramírez, S. Malato, M.I. Maldonado, Remediation of agro-food industry effluents by biotreatment combined with supported TiO2/H2O2 solar photocatalysis, Chemical Engineering Journal, Volume 273, 1 August 2015, Pages 205-213, ISSN 1385-8947, http://dx.doi.org/10.1016/j.cej.2015.03.060.

Index